RAPID REVISION NOTES
A LEVEL
BOTANY

RAPID REVISION NOTES
A LEVEL
BOTANY

BY JAN CASWELL BSc, Dip. Ed.,
Biology Dept.
Ebbw Vale Grammar School

General Editor – Professor A.J.B. Robertson,
Professor of Chemistry, King's College,
London

CELTIC
REVISION AIDS

Celtic Revision Aids
30-32, Gray's Inn Road,
London WC1X 8JL

© C.E.S.

First published 1982

ISBN 0-86305-113-8

All rights reserved

Printed and bound in Great Britain by
Cox & Wyman Ltd, Reading

GENERAL EDITOR'S FOREWORD

The Rapid Revision 'A' Level Series is designed for students preparing for G.C.E. 'A' Level, Scottish Highers, Intermediate University and similar examinations. They are comprehensive and may be used either on their own for revision or as a complement to set books and text books. The notes are organised to help students remember facts and to pin-point areas of difficulty. Practice questions are included at the end of each section with answers to the numerical problems at the end of the text. Where necessary, clearly worked through examples have been included in the text.

I am sure students will find these books very helpful.

A J B Robertson

M.A., Ph.D., D.Sc., C. Chem., F.R.S.C.

Professor of Chemistry,
King's College London,
University of London

Formerly Fellow of St John's College
  Cambridge

AUTHOR'S FOREWORD

This book, which specifically deals with Botany, is intended for students entering 'A' level examinations in Biology. Its purpose is to help and guide the student during preparation for the forthcoming examinations.

The syllabus has been divided into self-contained sections to remind the reader of the essential facts of the subject. At the end of each section are appropriate questions to test the knowledge acquired by the reader.

J. Caswell

Contents

1 The cell and its functions   1
2 Cellular types and distribution   11
3 Photosynthesis, nitrogen fixation and transpiration   23
4 Growth   35
5 Genetics and evolution   41
6 Plant communities, soil, pollution and conservation   51
7 Classification, Bacteria, Viruses, Fungi, Algae and Lichens   59
8 Bryophyta   77
9 Pteridophyta   83
10 Spermatophyta   91
11 Climbers, xeromorphs, halophytes and hydrophytes   117

1 The cell

The cell is the basic structural and functional unit of all living organisms. The plant cell consists of three main components.

The protoplast

This is the main living material of the cell. It is completely surrounded by differentially permeable membranes. Various sized particles or organelles are suspended in the cytoplasm, the area of protoplasm in the cell excluding the nucleus.

Nucleus

This is a spherical particle bounded by a nuclear membrane containing nucleic acid. There are two types of nucleic acid, deoxyribonucleic acid (DNA) and ribonucleic acid (RNA). Long strands of DNA occur in a network of chromatin. Each DNA molecule has a diameter of 2nm. DNA is a stable molecule composed of two parallel nucleotides arranged in a double helix. The nucleotides are held together by hydrogen bonds between the adjacent bases. There are ten base pairs in a double helix. Each nucleotide is composed of a base, either purine (adenine, or guanine) or pyrimidine (cytosine, or thymine), a pentose sugar and a phosphate. Bases attract one another and some form bonds which enable chains to fit together. The order of nucleotides in the chain determines the genetic information carried by the chain.

Within the nucleus and situated at a precise locus on a chromosome the nucleolus can be found. This is composed of ribonucleic acid and differs from DNA as each nucleotide contains the sugar ribose and the base uracil replaces thymine.

Ribosomes

Ribosomes are small spheroid particles in the cytoplasm, composed of RNA. Protein synthesis takes place in the ribosomes.

Protein synthesis

Each amino acid which makes a protein is first collected by a specific transfer-RNA in which form it is brought to the surface of a ribosome. The position of each amino acid in the protein chain depends on the order of the bases in a specific messenger-RNA coating the ribosome. Messenger-RNA is known to be complementary to a portion of a DNA molecule and the order of its bases is therefore determined by the order of the bases of that DNA. The bases attract one another and some form bonds which enable chains to fit together.

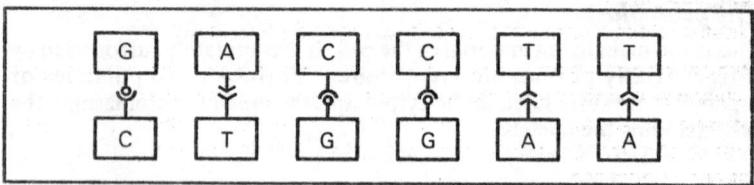

Hydrogen bonding accompanies base pairing and the chains become complementary. If one DNA locus consists of the bases AGC, the messenger-RNA formed on its surface would have the complementary bases UCG. The arrangement of three bases on the messenger-RNA forms a triplet code for a particular amino acid and the sequence of these three bases forms a codon. The messenger-RNA acts as a protein template. Peptide bonds (-NH-CO-) are formed between the amino acid and the polypeptide chain. This chain forms a globular or tertiary configuration and makes an **enzyme**. These enzymes have a specific catalytic property and regulate the rate of all chemical reactions in a cell.

Hydrolysing enzymes
These catalyse any break down of substrate by hydrolysis e.g. **Phosphorylases** catalyse the reversible conversion of starch to glucose phosphate. Oxidative phosphorylation is the coupling of phosphate with adenosine di-phosphate (ADP) to make adenosine tri-phosphate (ATP) utilising the energy released during respiration.

Oxidation and reduction enzymes
These enzymes catalyse oxidation in the presence of free oxygen e.g.

Oxidases discolour plant tissues in the presence of air. This occurs in broad beans, Vicia faba, when they turn red/black due to the oxidation of tyrosine to tyrosinase.

Dehydrogenases catalyse the oxidation of a substrate by removing hydrogen from it. The hydrogen may combine with molecular oxygen or with another substrate. Most oxidising enzymes are dehydrogenases e.g. succinic dehydrogenase.

Carboxylases
These catalyse the decarboxylation of pyruvic acid to acetaldehyde and carbon dioxide. They are involved with the liberation of carbon dioxide in cell respiration and with the assimilation of carbon dioxide in photosynthesis e.g. pyruvic carboxylase.

Decarboxylases
These convert amino acids to amines and carbon dioxide.

One enzyme catalyses one reaction or type of reaction. All enzymes form a chemical complex with the substrate. The substrate molecule fits onto the enzyme at one site – the active site. This enzyme substrate complex then undergoes internal re-arrangement that alters the substrate and releases the products.

Some enzymes are protein with a loosely bound non-proteinous molecule attached; this is a **co-enzyme**. All co-enzymes must be attached to the correct position on the protein. If the non-protein part is firmly bound it is a **conjugated protein** and the non-protein part **the prosthetic group**.

Co-enzymes play an important role in chemical reactions without being consumed in the process. e.g. cytochrome.

Enzyme inhibitors are substances which prevent the normal action of an enzyme. Sometimes inhibitors are removed by kinases or activators.

Nuclear membrane
The nucleus is separated from the cytoplasm by a nuclear membrane. This membrane is composed of two unit membranes 20 nm thick.
(1 nanometre = 10^{-9}m; 1 Angstrom (Å) = 10^{-10}m). The outer one is continuous with the endoplasmic reticulum and may bear ribosomes. The inner one is smooth and its inner surface provides a surface for the attachment of chromatin threads. During mitosis the nuclear membrane disappears allowing the nuclear and cytoplasmic material to mix freely. The nuclear membrane has large pores called annuli which allow large molecules to pass through it selectively.

Mitochondria
Mitochondria are rod-shaped structures up to 10 nm long and 1 nm wide. Each mitochondrion has a double lipo-protein membrane surrounding a dense matrix. These membranes provide strength, stability and flexibility. Outer and inner membranes are separated by a watery liquid. Cristae

extend from the inner membrane into the interior of each mitochondrion. The outside surface of the outer membrane and the inside surface of the inner membrane are sprinkled with elementary particles. These particles carry out the chemical activities of the mitochondrion. During mitochondrial oxidation the energy is stored in the form of special energy-rich phosphate bonds of adenosine triphosphate (ATP). ATP cannot be transported from cell to cell and so must be manufactured in each cell.

During respiratory metabolism the substrate molecule undergoes **dehydrogenation**. Nicotinamide adenine dinucleotide or NAD accepts two hydrogen atoms and is thereby reduced to NAD.2H. It then passes through a redox series of electron transfer and becomes oxidised to its original form by flavin adenine dinucleotide or FAD so that the chain can begin again.

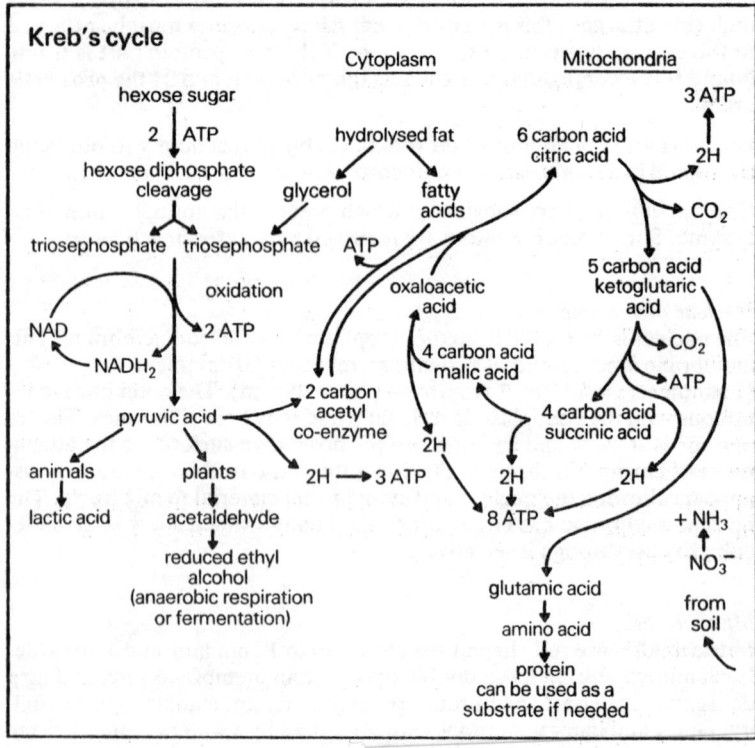

Where any free energy on the redox chain is excessive, ATP is reconstituted from ADP and inorganic phosphate.

Respiration
This is the breakdown of a substrate to liberate energy either aerobically or anaerobically.

The oxidation of a hexose sugar involves many enzyme-catalysed reactions. **Glycolysis** is the first stage of the chemical reaction whereby glucose is converted to pyruvic acid through a complex system of enzymes and co-enzymes. A small quantity of ATP is made available.

The respiratory quotient (RQ) is the amount of carbon dioxide produced divided by the amount of oxygen consumed in a given time.

$$C_6H_{12}O_6 + 6O_2 \longrightarrow 6CO_2 + 6H_2O + \text{Energy}$$

$$RQ = \frac{\text{Carbon dioxide produced}}{\text{Oxygen consumed}} = \frac{6}{6} = 1$$

Respiratory quotients are good indicators of the substrate being used.

Chloroplast
Plant cells contain cytoplasmic bodies called plastids. In green plants chloroplasts are found. They are disc-shaped bodies about 5-8 nm long and 1 nm wide. There are approximately 50 per cell. Each chloroplast is surrounded by a double membrane. Inside are layers of lamellae arranged parallel to one another. Each lamella is composed of two layers of lipoprotein with a space between them. The granal regions are thickened into grana. Surrounding the lamellae is a liquid stroma containing enzymes and starch grains. These chloroplasts absorb the radiant energy of the sun for photosynthesis.

Plastids become leucoplasts if they remain colourless and chromoplasts if they are coloured. Some plastids become the centres of starch formation and are called amyloplasts.

Cell wall

Plant cells have a rigid outer wall of non-living material which is secreted by the cytoplasm. The first layer is formed after cell division and is the middle lamella, at first composed of jelly-like pectin material and later of fibrils of crystalline, micelles of cellulose, polysaccharides and lignin. As the cell grows the cell contents exert pressure against the wall and so stretch it to accommodate the growth. As the cell grows the walls thicken and new layers are added to the old wall. When the cell reaches maturity it stops growing and rigid secondary material is added.

As the walls thicken the volume of cytoplasm decreases. In fibres the cell contents die and the walls occupy the whole cell. The walls are then used for support and in tracheids the empty cells are ideal for water conduction.

The secondary walls are perforated by many pits. In parenchyma cells the pits are simple cylinders of open spaces that run from the innermost part of the secondary wall to the outermost part of the primary wall. In dead cells e.g. tracheids, vessels and fibres, the cylindrical area is covered by a portion of secondary wall called a border and thus are known as bordered pits. The cell wall is perforated by holes through which pass cytoplasm, the plasmadesmata. The strands connect the cytoplasm of the other cells and materials are able to pass from one cell to another along these pathways. The cell wall acts as a skeleton giving strength and shape.

Vacuoles and membranes

A vacuole occurs in the cytoplasm. It is filled with cell sap composed of an aqueous solution of inorganic salts, organic molecules and waste products of cell metabolism. Each vacuole is surrounded by a membrane, the tonoplast. This is composed of protein and is differentially permeable, which allows the rapid passage of molecules of water and certain other substances, but permits only the slower passage of most dissolved solutes. The vacuole can exert a force on the cytoplasm and the cell wall.

The cell vacuole contains solutes. The water is less dilute than the water outside, therefore more water molecules will enter one cell than leave the vacuole. This rapid, unequal two-way diffusion of water through the membrane increases the volume of the vacuole and it presses against the cell wall. This is called turgor pressure and the cell becomes turgid.

The entry of water molecules stops when the turgor pressure of the cell reaches a point where the back pressure of the wall on the cell contents is great enough to prevent a further entry of water. This movement of water from a weak solution to a strong solution through a differentially permeable membrane is called **osmosis**. The osmotic concentration is a measure of the cell's maximum water absorbing tendency. Osmotic pressure is the pressure required to prevent net water entry into a cell of a given concentration.

When a cell is placed in a solution in which the concentration of water molecules is lower than that of the vacuole, the vacuole will lose water i.e. it is hypotonic (at a lower osmotic pressure). The vacuole will shrink so that the cytoplasm will withdraw from the cell wall producing **plasmolysis**.

The external solute concentration that produces incipient plasmolysis is a measure of the internal osmotic concentration. In a plasmolysed cell the space between the outer cell membrane and wall is occupied by the plasmolysing solution.

Division of a cell

Mitosis and Meiosis
Mitosis is the division of a cell into two identical cells.

Prophase
1. Division begins when the **chromosomes** in the nucleus become visible.
2. They become shorter and thicker and appear as paired threads.
3. Each **chromatid** of a pair is joined by a **centromere**.
4. The nuclear membrane disappears and the nuclear and cytoplasmic material mix freely.
5. A spindle-like arrangement extends from one pole of the cell to the other.

Metaphase
1. The spindle is complete.
2. The chromosomes arrange themselves along the equator of the spindle.
3. The chromosomes are attached to the spindle by the centromere.

Anaphase
1. The chromatids of each chromosome separate.
2. The chromatids begin to migrate towards the poles by the elongation of the spindle axis.
3. A spindle stem forms.
4. Gradually the chromatids approach the poles.

Telophase
1. The chromatids reach the poles.
2. The chromatids duplicate themselves to form chromosomes.
3. The chromosomes lose their dense appearance.

4. The nuclear material re-forms, bounded by a nuclear membrane.
5. Two daughter nuclei are formed.

Cleavage
1. Division of the rest of the cell is brought about by the formation of a groove on either side of the cell at the equator.
2. The protoplasm of the cell flows inwards on either side and extends towards the centre of the cell.
3. The flow continues until the new cell surfaces are formed.
4. Two daughter cells are formed and mitosis ends.

Meiosis
Meiosis differs from mitosis in that it produces four haploid cells. Two nuclear divisions occur.

Prophase
This is divided into five stages:

1. Leptotene:
a) The nucleolus disappears.
b) The chromosomes become long, thin threads in the nucleus.

2. Zygotene:
a) Chromosomes come together in homologous pairs.
b) They lie close together, centromere to centromere along their whole length.
c) The chromosomes become short, thick and twisted.
d) Bivalents are formed.

3. Pachytene:
a) The bivalents continue to shorten and thicken.
b) Each bivalent becomes a duplicate structure, becoming longitudinally split into two chromatids, but not at the centromeres.
c) Four twisted strands are formed as a result.

4. Diplotene:
a) The chromatids of each bivalent draw away from each other and form a tetrad.
b) Where the chromatids cross each other chiasmata are formed.
c) Stresses are set up, chromatids break and an exchange of chromatid material occurs. This is known as crossing over.
d) Terminalisation occurs when chiasmata between chromosomes move to the ends of chromosomes.

5. Diakinesis:

a) Chromosomes become short and thick.
b) Homologous chromosomes move apart.
c) The nuclear membrane disappears.
d) The spindle forms.

First meiotic division
1. Metaphase 1:
a) Each pair of homologous chromosomes come to lie at an equal distance from either side of the equatorial plane of the spindle.
b) Each is attached by its centromere.

2. Anaphase 1:
a) The centromeres move towards opposite poles of the spindle, drawing its two chromatids with it.
b) Terminalisation is completed.

3. Telophase 1:
Two new spindles are formed. They lie at right angles to the first spindle axis.

Second meiotic division
1. Metaphase 2:
a) The chromosomes become split into two chromatids.
b) Chromatids group themselves on the equator spindle.

2. Anaphase 2:
a) Centromeres divide.
b) Chromatids move away from each other.

3. Telophase 2:
a) The chromatids arrive at the polar areas.
b) Chromatids duplicate themselves.
c) Chromosomes lose their dense appearance.
d) The nuclear material re-forms bounded by a nuclear membrane.
e) Daughter nuclei are formed.

Four haploid cells are formed during meiosis. Reduction division has occurred.

Practice Questions

1. Name the three main regions of a plant cell.
2. Write a brief account of any three organelles.
3. Distinguish between DNA and RNA.
4. Where does protein synthesis occur and what parts do transfer-RNA and messenger-RNA play in the process?
5. What is a triplet code?
6. Distinguish between a hydrogen bond and a polypeptide bond.
7. What do you understand by the term enzyme? Give examples of the four groups of enzymes.
8. What is a co-enzyme?
9. Distinguish between anaerobic and aerobic respiration.
10. What is a respiratory substrate?
11. How is energy stored in a cell?
12. What is a hydrogen acceptor?
13. Explain glycolysis.
14. What do you understand by Krebs' cycle?
15. Describe the plant cell wall.
16. Explain the meaning of 'differentially permeable membranes'.
17. Define osmosis.
18. What do you understand by a) osmotic pressure and b) turgor pressure?
19. Distinguish between a flaccid cell and a turgid cell.
20. Define plasmolysis.
21. Explain the meaning of incipient plasmolysis.
22. What is the difference between mitosis and meiosis?
23. Describe in detail the process of mitosis.
24. Describe the first prophase of meiosis only.
25. What is the significance of meiosis?

2 Cellular types and distribution

Cell division is followed by the processes of cell elongation and cell differentiation. Cellular types arise from meristematic cells at the apex and develop through a process of differentiation.

Apical and lateral meristems
Plant growth is usually restricted to the meristems. These occur at the tips of organs e.g. root and stem. They are apical meristems. Lateral meristems arise from the cambium during secondary thickening.

A Tissue

A tissue is a collection of similar cells performing a similar function e.g. mechanical tissue.

Types of tissue in plants
Plant tissues are classified according to their functions.
1. Ground tissue e.g. parenchyma
2. Mechanical tissue e.g. collenchyma
3. Conducting tissue e.g. xylem and phloem
4. Protective tissue e.g. periderm

Types of cells

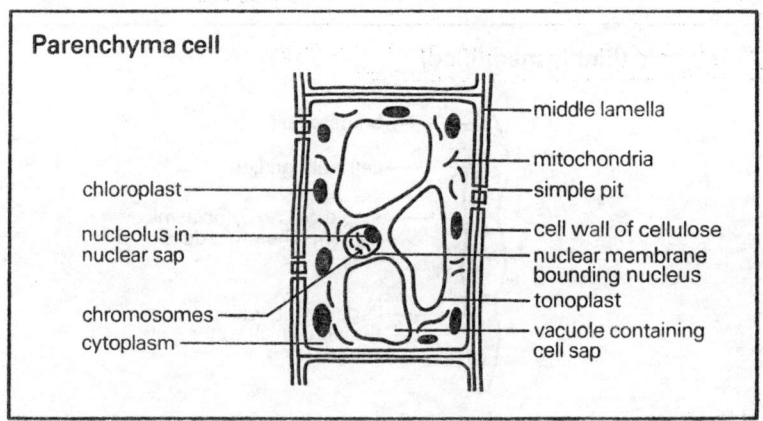

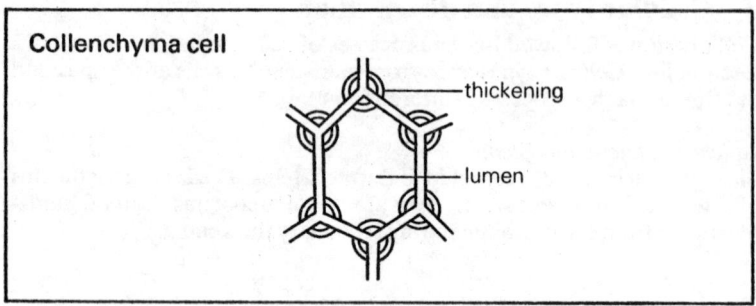

Collenchyma cells are found beneath the epidermis. They resemble the parenchymatous cells but have thickened walls of cellulose at the angles of the cells. They occur in organs that are still growing.

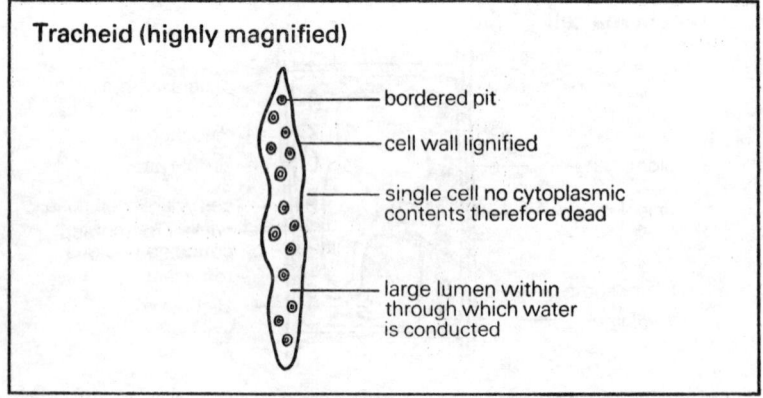

CELLULAR TYPES AND DISTRIBUTION

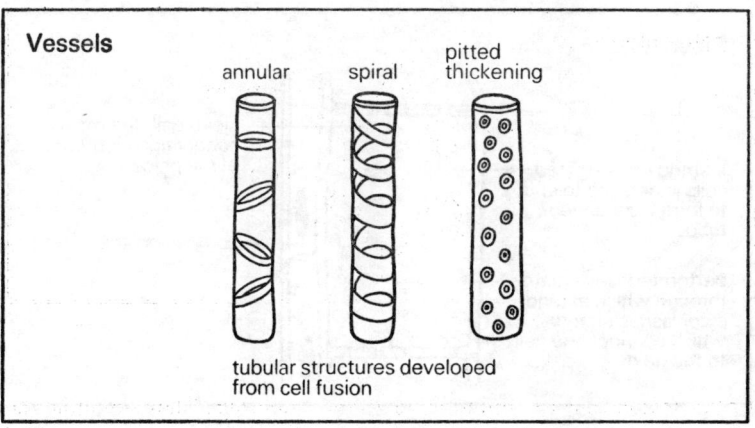

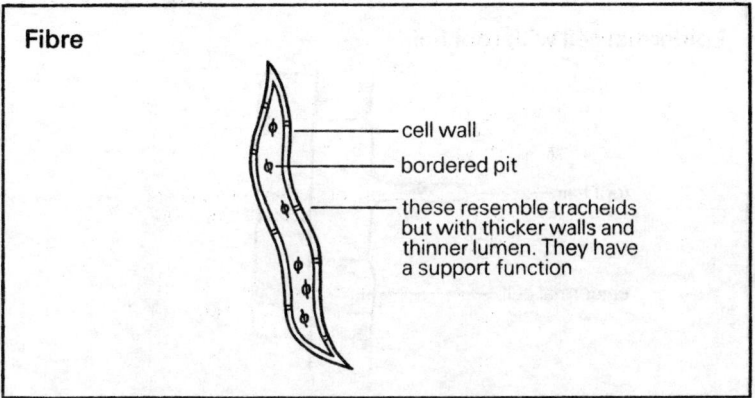

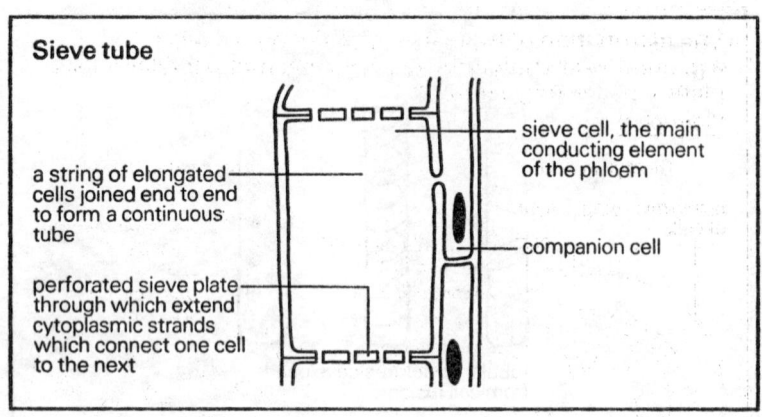

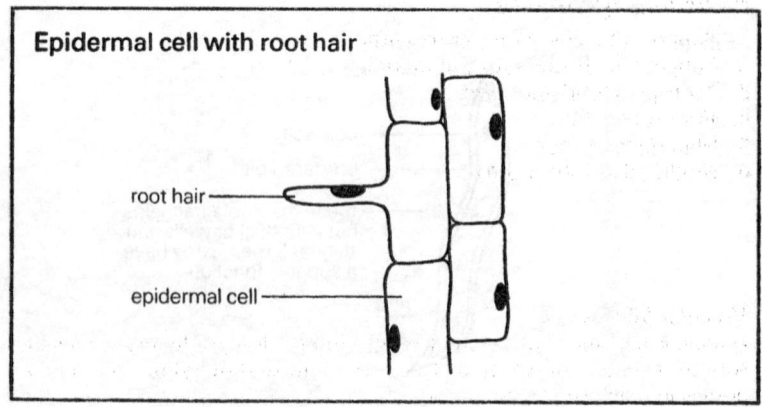

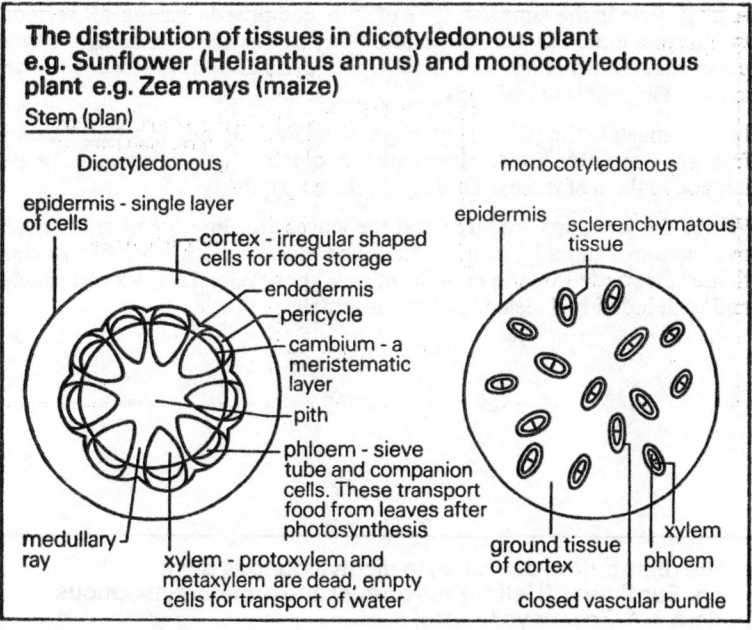

Functions of the stem
1. Support the leaves for photosynthesis.
2. Support the flowers for pollination.
3. Transport food and water.
4. For propagation.
5. Manufacture food.
6. Modified as food stores.

Vascular bundles
A stele is a cylinder of vascular tissue running longitudinally through the root and stem. The vascular tissue is composed of xylem, phloem and pericycle, enclosed by an endodermis.

A **protostele** is the simplest type of stele composed of a central core of xylem surrounded by a cylinder of phloem. Both are then bounded by a pericycle. There is no pith. Leaf traces arise as simple strands to the leaves. There are no leaf gaps.

In a **solenostele** the phloem, pericycle and endodermis together extend from the outside to line the inner surface of the xylem through small gaps left above the leaf traces. These gaps close up quickly.

Where the leaf gaps are large and the leaves are close together the gaps overlap longitudinally, causing the tube to form a mesh-work of vascular tissue surrounded by the endodermis and pericycle. This is a **dictyostele** and each individual stele is called a **meristele**.

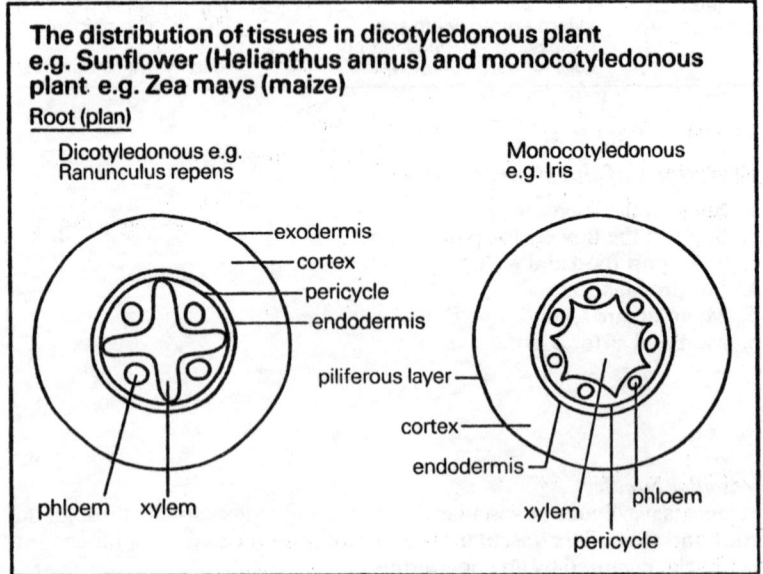

Functions of the root

1. Anchor plant in the ground.
2. Absorb water and mineral salts from the soil.
3. Transport food and water to the stem.
4. Modified for food storage.
5. For propagation.

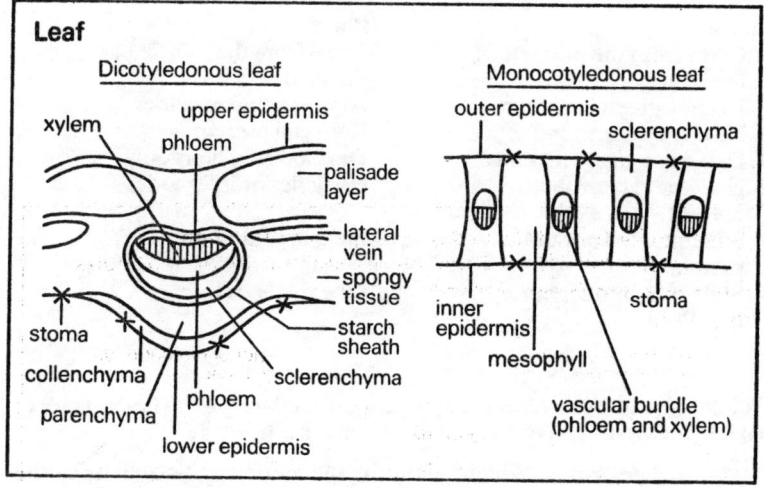

Functions of leaves

1. Manufacture food.
2. Diffusion of gases for respiration and photosynthesis.
3. Modified as organs of food storage and propagation.

Arrangement of vascular tissue in stem and root

Stems	**Roots**
Wind and animals bend the stems sideways. The arrangement of the vascular bundles or stele allows flexibility but gives strength around the outside.	Roots have to resist uprooting. The stele passes through the central core of the root to prevent them being pulled out of the soil.

Differences between stems and roots

Stems	**Roots**
1. Grow from the plumule of the seed	Grow from the radicle of the seed
2. Usually green in colour	Always white in colour
3. No cap present at apex	Root cap present
4. Develop leaves and buds	Develop root hairs only
5. Nodes and internodes present	No nodes or internodes
6. Stele situated around periphery	Stele in central position
7. Stele split into separate bundles	One single stele

Perennial plants grow from year to year. Secondary growth is necessary to enable the stems and roots of plants to increase in girth.

Secondary growth is brought about by the meristematic activity of the cambium, a lateral meristem.

The cambium ceases its activity in the winter but renews it in the spring. Every year the cambium forms a band of secondary tissue. Between each year an annual ring is found which separates the large elements of spring wood from the smaller elements of the winter wood.

Periderm
The phellogen arises in the outermost region of the cortex. Phellogen cells are produced by the division of both interior and exterior cells. The external cells are rectangular cells which become suberised, i.e. converted to cork, and die. They are part of an impermeable and insulating tissue.

CELLULAR TYPES AND DISTRIBUTION 19

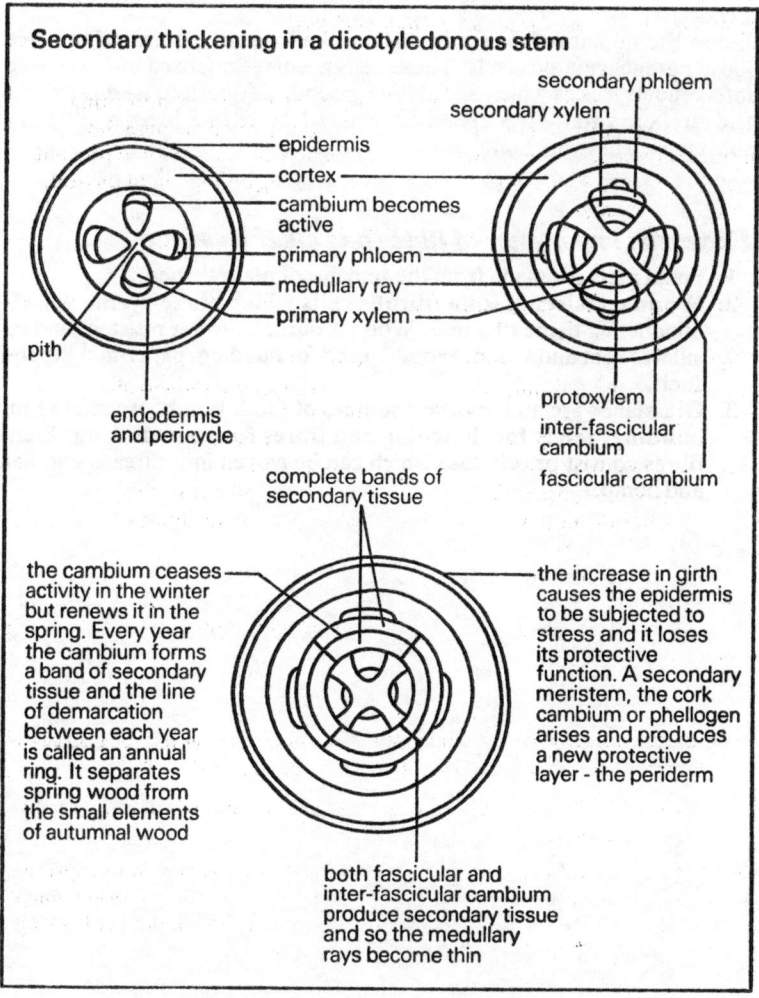

The inner cells become parenchymatous or collenchymatous and form what are known as the phelloderm.

    Phellogen + cork + phelloderm = Periderm

Lenticels
Below the stomata of the epidermis the phellogen of the lenticel produce loose parenchymatous cells. These cells become suberised but also have intercellular spaces which allow slow gaseous diffusion to take place. At the surface small slits are observed, filled with a brown powdery substance. These are lenticels.

Economic importance of fibres and other elements

1. Fossil fuels are made from the remains of plant tissues.
2. Wood consists of long fibrous cells which make up the water-conducting tissue of a tree. Wood is durable, water resistant and insulates heat and sound. Wood is used for building, paper-making and fuel.
3. Grasslands are an important sources of food. Bamboos are used for building, reeds for thatching and fibres for rope-making. Plant fibres consist of cellulose which can be woven into threads e.g. flax and hemp.

Practice Questions

1. What processes follow cell division?
2. What do you understand by meristematic cells and where are they found?
3. Make a large labelled diagram of a typical plant cell.
4. Distinguish between parenchyma, collenchyma and sclerenchyma.
5. Name the cells of the xylem and phloem and state the functions of each.
6. By means of diagrams only, distinguish between a monocotyledonous and a dicotyledonous stem.
7. State the functions of roots.
8. Tabulate the differences between stems and roots.
9. What is secondary thickening?
10. Describe secondary thickening in a stem.
11. By means of diagrams only, distinguish between a monocotyledonous and a dicotyledonous leaf.
12. What is a vascular bundle?
13. Distinguish between a protostele, siphonostele, solenostele and a dictyostele.
14. Distinguish between fascicular and interfascicular cambium.
15. What is a lenticel?

3 Photosynthesis, nitrogen fixation and transpiration

Photosynthesis

A leaf is composed of layers of photosynthetic cells, surrounded by a protective layer and supplied with a vein for conducting food and water throughout the plant. The surface of the leaf is perforated by stomata which form part of a system of interconnecting air ways. Stomata open and close in response to the turgor pressure of guard cells that surround them. The inner walls of the guard cells are thicker than the outer walls. When the guard cell is under great turgor pressure the weaker outer wall pushes out and carries the inner stronger wall with it, causing the pore to open. When the guard cell is flaccid, the thicker, elastic inner wall pulls the rest of the cell inward toward the pore area, thus closing it up.

Photosynthesis is the manufacture of organic compounds from water and carbon dioxide using the energy of the sun which is absorbed by chlorophyll.

$$6CO_2 + 6H_2O \xrightarrow[CHLOROPHYLL]{ENERGY} C_6H_{12}O_6 + 6O_2$$

The carbon dioxide reaches the palisade cells from the air through the stomata. The carbon dioxide passes through the pore and into the substomatal air chamber and diffuses throughout the intercellular spaces. On contact with the water surrounding the surface of the cell the carbon dioxide dissolves in the water to form carbonic acid, which is neutralised by the cations of the cell to form bicarbonate ions. Thus carbon dioxide can be used for photosynthesis. The carbon dioxide content of the substomatal cavity regulates the stomatal opening. If the carbon dioxide concentration is below 0.03% the guard cells become turgid and the stomata open. This can be brought about by the illumination of the guard cells, causing photosynthesis and thus using the carbon dioxide present.

The oxygen produced is liberated to the outside through the stomata. If the stomata are closed the oxygen can be used in respiration, producing carbon dioxide which is used in photosynthesis again.

Water is absorbed from the soil by roots through osmosis. It passes upwards through the xylem tissue in a transpiration stream. The water diffuses into the mesophyll cells in response to an osmotic gradient. Photosynthesis occurs in all green plants. If light energy, water and carbon dioxide are supplied to the cells and an efficient rate of transport of the products away from the cells occurs, an efficient rate of photosynthesis can be maintained.

The first stage in photosynthesis is the splitting of water, **photolysis**. Light energy is required. Visible light is absorbed by the chlorophyll pigments. The four main pigments are chlorophyll a, chlorophyll b, xanthophyll and carotene. These pigments become photochemically active. The activated pigment can remove an electron from the hydroxyl ion contained in the water molecule, and become deactivated in the process.

The free radical releases oxygen and forms free radicals of hydrogen. The hydrogen ions from the water and the electron attached to the pigment are transferred to a hydrogen acceptor which can take their reducing power to other reactions.

The split products of water recombine to form water. This reaction releases energy. The chloroplast containing the chlorophyll causes the reaction to make energy rich ATP from a molecule of ADP and inorganic phosphate.

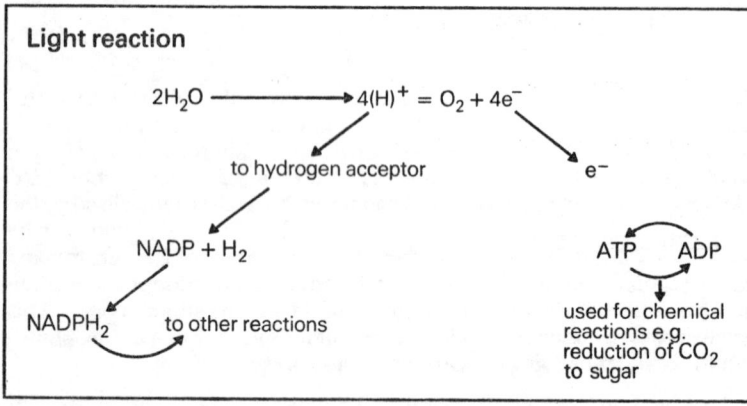

The light energy produces ATP through a complex series of reactions called **photophosphorylation**. The production of carbon dioxide occurs in darkness.

The first stable product of photosynthesis is 3-phosphoglyceric acid, (PGA). This is formed from the combination of carbon dioxide with ribulose diphosphate to form an unstable 6-carbon compound which decomposes into two PGA molecules.

The phosphoglyceric acid is reduced to an aldehyde by the reducing power of NADPH and the energy of ATP. These are both generated by the photolysis of water.

The phosphoglyceraldehyde, a sugar phosphate, has only 3 carbon atoms, while the smallest sugar has six. The phosphoglyceraldehyde molecules combine to yield a phosphorylated hexose i.e. fructose diphosphate which can be de-phosphorylated to form free hexose sugar.

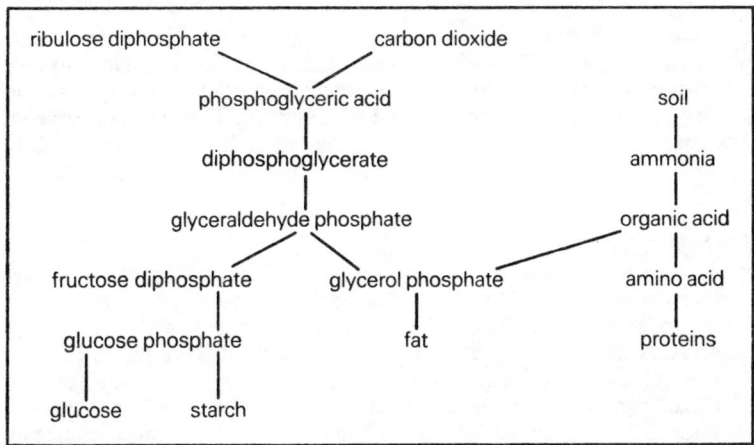

Oxygen is produced as a by-product of photosynthesis. The first step involves the ionisation of water into $[H^+]$ and $[OH^-]$. An electron is removed from the hydroxyl ion to produce an hydroxyl free radical [OH]. The fusion of two [OH] radicals occurs to form a peroxide which would decompose to release oxygen.

The concept of limiting factors
The rate at which photosynthesis occurs depends on whether any factor is in short supply. Light becomes a limiting factor at night and carbon dioxide by day. The temperature may be too cold and therefore the reaction slows down, while too hot a temperature may destroy the enzymes. Lack of magnesium in the soil prevents the formation of chlorophyll and the closing of stomata may reduce the flow of carbon dioxide into the leaf.

Compensation point is reached when the amount of carbohydrate manufactured in photosynthesis is equal to the amount of carbohydrate broken down in respiration.

In plants, sugars produced in photosynthesis are converted into starch which accumulates in the chloroplasts. This enables more glucose to be formed as an equilibrium is avoided. At night the starch is hydrolysed and translocated to the growing regions and storage organs, where it is stored in cytoplasmic plastids called amyloplasts.

Proteins
Proteins contain the elements carbon, hydrogen, oxygen, nitrogen, phosphorus and sulphur. They form large molecules and are organic. Proteins are composed of chains of amino acids which contain both an amine ($-NH_2$) group and a carboxyl ($-COOH$) group. They are amphoteric. The condensation of carboxyl groups and amino acids form a peptide link; $-CO-NH$. Amino acids form polypeptide chains. The test for protein is the Biuret test.

Carbohydrates
1. Monosaccharides have the general formula $C_nH_{2n}O_n$ e.g. glucose, fructose and galactose. They are reducing sugars and may be either triose, pentose or hexose. The test for monosaccharide is with Fehling's A and B, when a brick-red precipitate of cuprous oxide is formed.
2. Disaccharides are formed from the condensation of two hexoses with the elimination of water e.g. sucrose, maltose and lactose. To test a disaccharide sugar it must first be reduced to the monosaccharide state and then tested with Fehling's A and B. A glycoside bond joins the two simple sugars.
3. Polysaccharides are important structural and storage compounds. They have a high molecular weight. They are formed by the condensation of many hexose units. Starch is formed from the condensation of glucose molecules to form a helical chain, while cellulose is formed from the condensation of glucose molecules to form a straight chain. Iodine is used as a test for starch. Cellulose must be treated with strong sulphuric acid before it will give a positive reaction with iodine.

Lipids
Fats and oils are esters produced from the condensation of glycerol and a fatty acid. Fats are solids whereas oils are liquid. Fatty acids form long chains of $-CH_2$ groups with a terminal methyl or carboxyl group e.g. palmitic acid $CH_3(CH_2)_{14}COOH$. Waxes e.g. beeswax, are esters of fatty acids with alcohols of high molecular weight.

Phospholipids are mixed lipids of glycerol, fatty acids and phosphoric acid e.g. lecithin. Plant steroids are sitosterol and stigmasterol which are found in the seed embryos.

Essential Minerals

An essential element is one without which the plant cannot complete its life cycle. To discover which elements are required by a plant water culture experiments can be performed.

All mineral elements except nitrogen are derived from the soil, which has been formed as a result of the corrosion of the parent rock. Nitrogen is obtained from the atmosphere, mainly through the fixation of nitrogen by bacteria. The main absorbing cells are in the root apices. The mineral elements must pass through a differentially permeable membrane of a cell. The membrane has numerous specific absorption sites through which particular ions enter. Ion absorption involves an expenditure of energy, for when aerobic respiration ceases the rate of ion absorption decreases.

The ions in the plant cells may be different from those in the surrounding soil water because the plant seems able to select some ions and reject others. A required ion is collected at the surface of the plasma membrane by a carrier molecule and transported across. When K^+ is taken into the cell and an anion OH^- is also picked up to keep the ionic balance. Similarly if NO_3^- is needed a cation H^+ is taken up. **Active absorption** takes place and the balance is kept by an exchange of cations and anions.

Some elements can be accumulated in the cell against an electrochemical gradient, this requires an expenditure of energy. If aerobic respiration is inhibited, the accumulated ions readily leak from the cell. Potassium and lithium enter at the same site and mutually inhibit each other's entry, whereas sodium enters at another site.

When a root secretes ions to the outside medium in exchange for ions absorbed the pH of the surrounding medium alters, e.g. the secretion of H^+ makes the external medium more acid.

Elements essential but in small amounts are called trace elements.

Elements	Use
Nitrogen (N)	This is an essential element of proteins which are needed for growth. Without nitrogen plants remain stunted and undeveloped.

Phosphorus (P) — Present in structural components of nucleic acids. The phospholipids play a part in the structure of membranes.
Energy transfer ATP is composed of three phosphates coupled to a ring structure. Two phosphate groups differ from the third as they produce far more energy during hydrolysis. ATP has two energy rich bonds $A - P \sim P \sim P$. Cleavage of ATP to ADP and P releases large amounts of energy.
The transformation of ADP back to ATP by the release of energy is an oxidative reaction. Lack of phosphorus causes a lack of new genetic material and no membranes.

Sulphur (S) — This is an essential element of amino acids. It has a structural function as it is found in glutathione and thioctic acid which plays a part in oxidative-reduction reactions.

Calcium (Ca) — This element helps to form the middle lamella of all cell walls. It stiffens and makes rigid a semi-fluid structure called pectin.

Magnesium (Mg) — This element forms the centre of the chlorophyll molecule, attached to four pyrrole rings either by direct co-valent bonds or by secondary valencies. Deficiency causes chlorosis.

Potassium (K), Iron (Fe), Copper (Cu), Manganese (Mn), Zinc (Zn), Molybdenum (Mo), Boron (B), and Chloride (Cl). — These elements play an essential catalytic role. Many enzymes consist of specific proteins to which are attached special chemical elements called **prosthetic groups** or **coenzymes**. These groups consist of a metallic element e.g. iron is found in cytochromes, peroxidase and catalase. The central iron atom is connected to four pyrrole rings joined into a large cyclic structure. Iron functions by its oxidation and reduction qualities ($Fe^{3+} + 2e^- \rightarrow Fe^{2+}$). Similarly with copper ($Cu^{2+} + 2e \rightarrow Cu^+$). Potassium activates enzymes. Chloride stimulates photosynthetic phosphorylation. Boron is involved in the rate of sugar movement. Molybdenum is involved with the enzyme nitrate reductase which reduces nitrate to ammonia.

Nitrogen fixation

Nitrogen occurs in the atmosphere as a stable nitrogen molecule. This nitrogen must be split before reduction of the nitrogen to ammonia can occur. This can be done by certain bacteria that live on organic matter in the soil, e.g. Azotobacter, an aerobic form, Clostridium, an anaerobic form, and the bacteria Rhizobium which lives in the nodules of leguminous plants, e.g. clover, Trifolium repens.

The association of Rhizobium with Trifolium repens, the host, is a symbiotic relationship. Once inside the host cell, the bacterium causes a prodigious growth or nodule. Each nodule contains a red pigment called leghaemoglobin which is involved in nitrogen fixation. The activation of nitrogen is brought about by nitrogenase which is inhibited by molecular hydrogen, and an electron transport agent called ferredoxin.

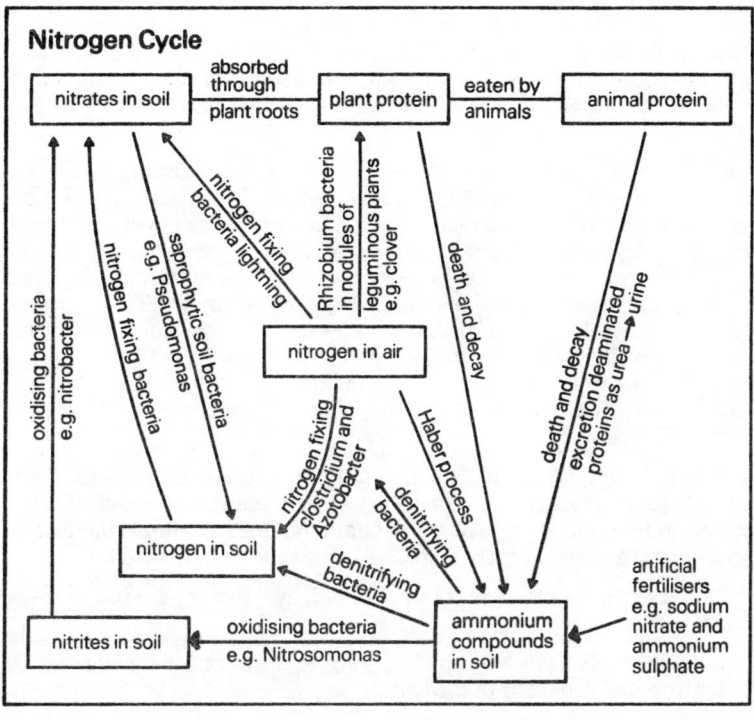

Most plants absorb and utilise nitrogen in the form of nitrate. Nitrate reductase reduces nitrate to ammonia by means of reduced respiratory carriers such as NADP.

Plant water and Transpiration

Plant cells are composed of 90% water. They therefore require large quantities of water, which they absorb from the medium in which they grow. Plants also lose large quantities of water to the atmosphere. The evaporation of water from the aerial parts of the plant is called **transpiration.**

The spongy mesophyll cells evaporate large quantities of water into the intercellular spaces. Diffusion of water vapour causes this water to find its way to the stomata and to the outside environment. If there is low humidity and turbulent air surrounding the plant, water will evaporate from the surface of the leaves. Thus, water from the intercellular spaces moves towards the sub-stomatal chamber and passes along a diffusion gradient from the leaf to the atmosphere, increasing the transpiration rate.

If humidity is high, water accumulates in the leaves, eventually oozing on to the surface through the stomata and dripping from the leaves. This is guttation. The fuschia and many other plants which exist in conditions of high humidity have hydathodes, special structures which actively secrete water. The closing of the stomata decreases the rate of transpiration and prevents the desiccation of the plant. It is the carbon dioxide content of the substomatal chamber that regulates the stomatal aperture. If the carbon dioxide concentration falls below 0.3% the guard cells become turgid and the stomata open. This can be brought about by the illumination of the guard cell, causing photosynthesis and decreasing the carbon dioxide content.

Xerophytes are plants that are adapted to life in arid conditions. They possess fewer stomata than mesophytes, sunken stomata and a highly cutinised leaf and stem. All these characteristics minimise the loss of water by restricting air turbulence.

All water lost by leaves comes from the soil, absorbed by the roots during osmosis.

Several theories have been put forward to account for the passage of water from the root cells to the leaves.

Transpiration cohesion theory
The evaporation of water from the mesophyll cells causes them to lose turgor so that a water potential gradient is set up. This gradient pulls up the water which adheres to the dead xylem tissue which have thick rigid walls and narrow lumina.

The mechanical pull of water results in a mass lifting of a cohesive column of water up through the mesophyll cells which are placed under a condition of tension. The energy needed for the rise of water in the xylem comes from the evaporation of the water from the leaf and is independent of the vital activities of the cells.

Root pressure theory
Root pressure causes an upward movement of water in the xylem tissue. This pressure is sufficient to move the water several feet upwards. It is thought that a combination of the transpiration cohesion theory, root pressure and capillarity enable the water to reach the leaves of the tallest trees.

Transpiration provides a cooling effect for the plant and also enables the salts absorbed by the roots to find their way into the xylem tissue. These salts are transported along with the water in the transpiration stream.

Translocation in the phloem
Many substances move in the phloem. The rates of movement exceed those caused by diffusion. Simultaneous movement of substances up and down the phloem can occur, but not in the same sieve tube. The contents of the sieve tubes move as a mass in response to hydraulic pressure gradients. The photosynthetic cells of the leaf generate a high osmotic pressure because of the high sugar content. This will cause high pressure in adjacent cells. Movement downward then occurs in response to the pressure gradient from above to below, resulting in a large scale transportation of sugar from the top to the bottom of the plant. This is a theory put forward by Munch but the actual mechanism is not yet clear.

Experiments

Photosynthesis
1. To show that oxygen is produced during photosynthesis.
2. To test a leaf for starch.

3. The conditions necessary for photosynthesis: carbon dioxide, light and chlorophyll.
4. To determine the relative rates of photosynthesis under varying conditions.
5. Absorption spectrum experiment.
6. The separation of the chlorophyll pigments by paper and column chromatography.
7. Water culture experiments.
8. Food tests.

Transpiration
1. To show that xylem tissues transport water.
2. To demonstrate root pressure.
3. To demonstrate that a plant transpires.
4. To show that water is given off from the undersurfaces of leaves.
5. The comparison of transpiration with a physical phenomenon.
6. To measure the rate of transpiration or the amount of water absorbed using a potometer.
7. To examine the conditions affecting the rate of transpiration.

Practice Questions

1. What is photosynthesis?
2. How is the leaf adapted for photosynthesis?
3. Give an account of the opening and closing of the stomata.
4. Write a brief account of chlorophyll pigments.
5. Describe the first stage of photosynthesis.
6. Name the first stable product of photosynthesis and what is the fate of this substance?
7. Distinguish between oxidative phosphorylation and photophosphorylation.
8. What do you understand by the concept of limiting factors?
9. What is the compensation point?
10. Write an account of either a) proteins, b) fats or c) carbohydrates.
11. Explain the following a) proteins are amphoteric, b) condensation, c) a peptide link, d) a reducing sugar, e) a hexose sugar, f) a glycoside bond and g) an ester.
12. What is the function of a) Fehling's solution and b) the Biuret test?
13. Name three essential elements and explain their importance to the plant.
14. Write an account of ion absorption.
15. What is a symbiotic relationship? Give an example to explain your answer.
16. What are leghaemoglobin and ferredoxin?
17. Name two nitrogen fixing bacteria.
18. Give an account of the nitrogen cycle.
19. What is transpiration?
20. What conditions increase the rate of transpiration?
21. Distinguish between a xerophyte and a mesophyte.
22. How does the water reach the topmost leaves of a tree from the roots?
23. What is the significance of transpiration?
24. What substances are transported in the phloem?
25. How are these substances transported?

4 Growth

Growth involves the formation of new protoplasm from basic raw materials. Once cells have reached maturity they divide, extend and differentiate into new organs. Measurements of growth are estimated in dry weight.

The seed contains an embryo plant surrounded and protected by a seed coat and supplied with a source of stored food.

The plant embryo contains a young root growing tip and a shoot growing tip. Cotyledons or embryonic leaves are attached laterally near the midpoint of the root and shoot. Cotyledons are thin, elongated and leaf-like. They help to digest the stored food in the endosperm tissue and can expand into leaf-like photosynthetic organs.

In other cases they are fleshy storage organs above or below ground that have absorbed the endosperm before maturation of the seed. These cotyledons do not become photosynthetic.

Conditions necessary for germination

1. Water
2. Suitable temperature
3. Adequate supply of oxygen

The seed absorbs large quantities of water causing the growing point to be stimulated into mitotic activity. The root develops first and is then followed by the growth activity of the shoot growing point. At both apices growth is due to the formation of new cells by the meristematic tissue in the growing regions. Growth is followed by elongation and differentiation of these cells.

In the root cell division, elongation and differentiation each occur in well-defined regions that overlap one another. Roots grow downwards and the growing point is protected by a root cap. This cap is produced by divisions of the meristem and is continuously flaking off and being replaced. The typical growth curve of all organs, plants and populations is illustrated by a Sigmoid growth curve.

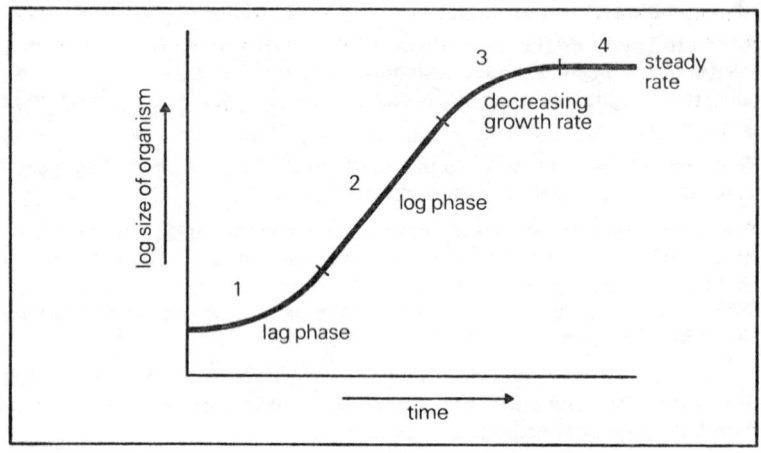

1. Initial lag period during which internal changes occur that are preparatory to growth.
2. A phase of increasing growth rate. The log of growth rate, when plotted against time, is a straight line and is referred to as the log period of growth.
3. The growth rate gradually diminishes.
4. The organism reaches maturity and growth ceases.
   The rate of growth of a plant depends on both genetic constitution and various environmental factors.

Hormones

Hormones are needed in small quantities and are produced in one part of an organism and transported to another part where they produce some special effect e.g. auxins, gibberellins and kinins.

Auxins are produced in growing apices of stems and roots. They diffuse from the apex to the zone of elongation where they are required for elongation. If the growing tip is removed growth ceases, but if the tip is replaced growth continues. The auxin can be collected on a block of gelatin which can then be substituted for the growing tip. The auxin is B-indole acetic acid (1AA). If portions of a growing stem are placed in a petri dish containing sucrose and mineral salts growth will be slow. If 1AA is added growth is enhanced and the effect is directly proportional to the concentration of the 1AA added.

If auxin is added to an intact stem the plant produces no increase of growth. This is because the stem is normally saturated with auxin produced by its own tip.

In roots, the addition of auxin usually results in growth inhibition and when the root tip is completely removed growth is intensified. The root works under conditions of sufficient auxin supply. Roots are more sensitive to auxin than stems. The growth of the root is promoted by concentrations of auxin lower than those that promote growth of the stem and is inhibited by concentrations lower than those that inhibit the growth of the stem.

The amount of auxin in a plant tissue can be determined by extracting it with a solvent, di-ethyl ether, at 0°C and gently shaking it for two hours. Application of the extract to the same tissue will cause it to respond quantitatively to the auxin contained in it. This is a bioassay technique.

Tropisms
Curvatures towards or away from light or gravity are produced by the unilateral application of auxin to plants and are known as tropisms. Such growth curvatures are caused by the asymmetrical distribution of auxin, e.g. Oat coleoptiles (Avena fatua) subjected to a low intensity of unilateral light will curve towards the light. This is **phototropism.** The side near the light has had its growth depressed by the light, while the growth on the side away from the light has been accelerated. Light acts in producing curvatures by affecting the distribution of auxin in the tissue, thus auxin concentration controls growth. (It is thought that light destroys 1AA by activating a 1AA oxidase enzyme). If a tip of a unilaterally exposed coleoptile is removed and the amount of auxin is measured in the two sides, the side away from the light will have approximately twice as much auxin in it compared to the side towards the light.

Geotropism
A stem laid on its side will accumulate more auxin on the lower surface than on the upper surface. Growth will be accelerated on the lower surface and a growth curvature upwards will occur.

The growth downward of the root is due to the difference of auxin sensitivity. In prostrate roots as in prostrate stems auxin accumulates below, but since in roots the auxin concentration is already at its maximum, this greater concentration of auxin on the lower side leads to depressed growth, thus a downward growth curvature occurs.

Importance of auxins
1. They promote cell division with cytokinins.
2. They determine the growth of the apical or lateral buds.
3. They regulate the fall of leaves and fruits from plants.

Synthetic auxins
Synthetic auxins are indolebutyric acid and naphthaleneacetic acid. If Naphthaleneacetic acid is placed on a tomato stigma it produces an artificial fruit without viable seeds. Auxin stimulates the development of the ovary wall which suggests that pollination results in a natural increase in auxin concentration in the ovary wall. Pollen contains a little auxin and its function is therefore to activate the formation of auxin. Artificial auxins can inhibit the growth of certain kinds of cells. Some have differential toxicity towards different plants. Thus 2.4.D applied to a mixed population of plants may kill one type and leave another unaffected.

Gibberellins
The application of gibberellic acid to plants has the effect of greatly elongating stems and inhibiting the growth of the main root, occasionally causing a diminution of the leaf area. It promptly stimulates flowering in a class of plants called long day plants.

Application of gibberellins to seedless grape clusters results in the retention and development of a greater number of larger grapes.

Growth inhibitors
Inhibitory substances are produced during the growing season and accumulate in the bud, causing it to be dormant. Dormancy is broken by the destruction of the inhibitor when the plant is exposed to a sufficiently long cold period.

Vernalisation
Certain plants will not flower until a minimum of vegetative growth has been achieved. Some flowers will not flower unless subjected to low temperatures for a long time, a process called vernalisation.

Circumnutation
The apex of a growing tip grows in a spiral due to the continuous change in

position of the most rapidly growing region of the tip. This is clearly seen in the tendrils of the everlasting pea, Lathyrus sylvestris.

Nastic movements
A movement made by a plant in response to a non-directional stimulus e.g. temperature, light, humidity and light intensity may cause the opening and closing of flowers. Sudden changes in turgor pressure may bring about the response of leaf movements in Mimosa pudica.

Photoperiodism
The response of plants to the length of day is photoperiodism. Some plants are short day and others long day plants. The leaves are the receptors of the photoperiodic stimulus. The stimulus is transmitted through the plant, via the phloem, to the terminal bud where it influences the meristem to produce flowers. The photoreceptor substance is phytochrome.

Most plants respond to the length of uninterrupted darkness. This makes a short day plant a long night plant which will require an uninterrupted dark period of a certain length of time before it will flower. A long day plant means that it is a short night plant and will only flower if the dark period is not longer than a certain critical minimum. If the dark period is interrupted, the flowering can be delayed or accelerated.

The two types of plants possess the same kind of photoperiodic mechanism but they work in reverse. The same kind of light will therefore inhibit the flowering of short day plants yet will promote the flowering of long day plants.

The phytochrome occurs in two forms 1) r-phytochrome (Pr or P_{660}) which absorbs red light and 2) fr-phytochrome (Pfr or P_{730}) which absorbs far-red light. When one form of the phytochrome absorbs light it is converted into the other form, while in the dark Pfr is slowly changed back into Pr. Pfr is an active form and initiates enzyme action. Pr is inactive.

In short day plants the absorption of red light in the middle of a long dark period prevents flowering, while the absorption of far-red light after the red light leads to the promotion of flowering.

Practice Questions

1. What is growth?
2. Describe a plant embryo.
3. What are cotyledons?
4. Enumerate the conditions necessary for germination.
5. What do you understand by a sigmoid growth curve?
6. On what does the rate of growth depend?
7. What is a hormone?
8. Give an account of B-indole acetic acid.
9. What do you understand by a bioassay technique?
10. What is a tropic response?
11. Distinguish between a phototropic response and a geotropic response.
12. Explain how the difference in auxin sensitivity affects the growth of a prostrate stem and root.
13. Distinguish between an auxin and a gibberellin.
14. State the difference between a tropic movement and a nastic movement.
15. Write an account of photoperiodism.

5 Genetics and Evolution

Genetics

Genetics is the study of inheritance. Mendelism is the theory related to the way in which parental characteristics are distributed in the offspring. Mendel studied the garden pea, Pisum sativum. The plants he selected showed contrasting characters e.g. tall and short plants, round and wrinkled seeds, yellow and green cotyledons.

Monohybrid inheritance is the inheritance of a pair of contrasting characters e.g. when a pure-bred tall plant is crossed with a pure-bred short plant:

Let germinal unit for tall be T.   A pure bred tall plant is TT.
Let germinal unit for short be t. A pure bred short plant is tt.

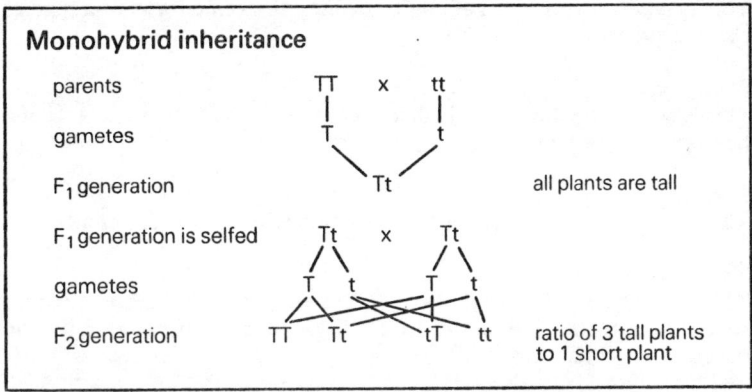

Phenotypes i.e. the general appearance of the plants are 3 tall plants and 1 short plant.

Genotypes i.e. the genetic compositions of the individuals are
1 TT : 2Tt : 1tt

Mendel's first law of inheritance states that of a pair of contrasting characters only one can be represented by a single gamete. This is the law of segregation.

Alleles (Allelomorphs)
Two genes are said to be alleles of one another when they occupy the same locus on homologous chromosomes.

Dominant
The gene whose characteristic appears in the heterozygous phenotype is said to be dominant e.g. T (Tall) is dominant to t (short) in monohybrid inheritance in the above example.

Recessive
The gene whose characteristic does not appear in the heterozygous phenotype is said to be recessive e.g. t (short) is recessive to T (tall) in the monohybrid inheritance as above.

Homozygous
This is where two identical alleles occupy the same locus of a pair of chromosomes.

Heterozygous
This is where two unidentical alleles occur on the corresponding loci of a pair of chromosomes.

Back cross
This is a cross of the hybrid or F_1 generation with one of the parents.

Test cross
This is the cross of the hybrid or F_1 generation with the homozygous recessive parent. This method distinguishes the homozygous and heterozygous offspring.

Incomplete dominance
In Snapdragons, Antirrhinum majus, the white character is not completely dominated by the red character and so pink flowers occur.

Let germinal unit for red flowers be R.   A pure red flower is RR.

Let germinal unit for white be r.   A pure bred white flower is rr.

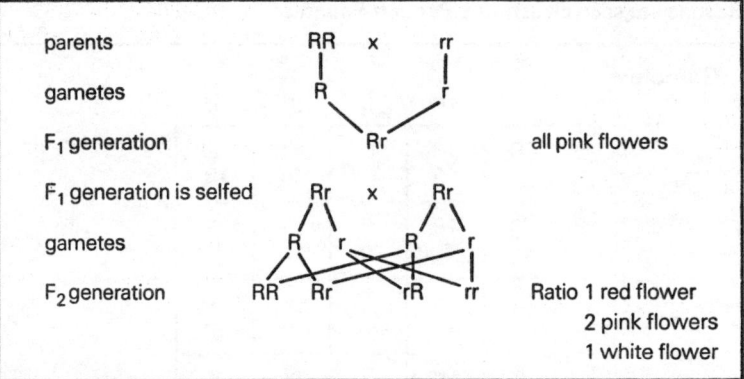

Genetic constitution of the individuals are 1RR : 2Rr : 1rr

Dihybrid inheritance

This is the inheritance of two pairs of contrasting characters. Mendel crossed pure-breeding tall, round seeds and pure-breeding short, wrinkled seeds.

Let germinal unit for tall be T.

Let germinal unit for short be t.

Let germinal unit for round be R.

Let germinal unit for wrinkled be r.

Therefore a pure-bred tall, round plant is TTRR.

Therefore a pure-bred short, wrinkled plant is ttrr.

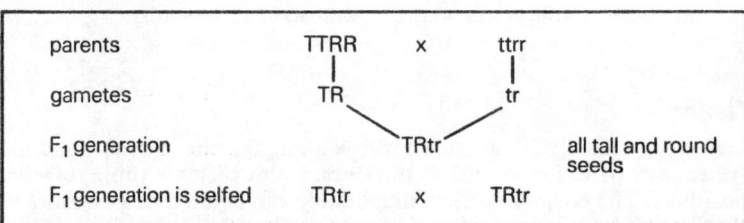

Results are seen clearly in a Punnett Square:

Gametes				
	TR	Tr	tR	tr
TR	TR TR	Tr TR	tR TR	tr TR
Tr	TR Tr	Tr Tr	tR Tr	tr Tr
tR	TR tR	Tr tR	tR tR	tr tR
tr	TR tr	Tr tr	tR tr	tr tr

F_2 generation:- 9 tall round; 3 tall wrinkled; 3 short round; 1 short wrinkled.

Mendel's second law of inheritance states that each one of a pair of contrasting characters may combine with either one of another pair. This is the law of independent assortment.

Linkage

Occasionally genes on the same chromosome may not segregate and are transmitted from one generation to another as a separate unit. These are linked genes. The nearer these genes are together the less chance there is that they will become separated in meiosis during the crossing over phase. All these inseparable genes on the chromosome form a linkage group.

Genes

Genes are arranged in a linear fashion along the chromosome and the frequency of recombination between them reflects their relative positions. The frequency of recombination can be used as a measure of the distance between the pairs of genes and, therefore, a genetic map can be made. Gene loci which give a low frequency of recombination would be closer together than those which give a high frequency of recombination. One unit on the map would be equivalent to 1% of recombination or crossing over, providing a means of determining the distances between genes on a chromosome.

Lethal genes
Some genes are considered to be lethal as their occurrence causes the death of the organism. The absence of a gene for chlorophyll production occurs in some cereal plants. If the gene causes heterozygous organisms to die it is a dominant lethal gene, but if only homozygous individuals die, it is a recessive lethal gene.

Chromosomes

These are thread-like structures found in the nucleus of all cells. They occur in homologous pairs. Along the length of each chromosome is a linear series of genes which carry the genetic information of each individual and which are capable of replication. Deoxyribonucleic acid (DNA) carries the genetic codes which determine the type of protein manufactured.

All nucleic acids are composed of nucleotides linked together, end to end. (See earlier).

Variations

Variations are differences that occur between individuals of the same species. Variations may be due to differences that occur in the environment during development e.g. food, light, temperature, water, or genetic differences which arise by the recombination of genes from mutations or re-assortments.

Continuous variation is controlled by several genes and influenced by environmental factors e.g. the number of seeds and size of fruit are not inherited.

Discontinuous variation is controlled by one or a few genes. The environment has no effect on them and they are heritable e.g. tall and dwarf peas.

Sources of variation

Mutation

This is a sudden change in the chromosomes. A change in the structure of DNA produces a gene mutation e.g. copper colour leaves and the weeping habit. Mutations occur during meiosis, especially in the diplotene phase (crossing over).

Gene and chromosomal mutations

These occur by:
1. Deletion when part of the chromosome is lost because of the loss of a nucleotide.
2. Inversion when the chromosome is turned around.
3. Translocation when the chromosome or part of the chromosome becomes attached to another chromosome.
4. Duplication when the chromosome replicates itself.
5. Insertion when the chromosome gains a nucleotide.
6. Substitution when the wrong nucleotide is placed in position.

In all the above examples the gene sequence has been altered. This alteration may be beneficial or detrimental to the individual, but most mutations are recessive and harmless. Mutations are the raw material of evolution.

Mutations can be increased by X-rays, gamma rays, ultra violet rays, mustard gas, nicotine and various drugs.

Evolution

Variation within a species means that some individuals become better suited to their environment than others of the same species. Thus a process of natural selection occurs and plants have developed from earlier forms. A theory of organic evolution has arisen. Palaeobotany is the study of the record of past plant life as shown by fossils.

A fossil is the remains of a plant, hardened and preserved in the strata. Volcanic upheavals may cause large areas of plants to become covered with ash, thus entombing the plants and leading, eventually, to their fossilisation. Fragments of vegetation are deposited in seas, lakes or ponds along with dust and other particles. Plant bodies that are not decayed become covered with these particles and eventually become fossilised.

Plant fossils are usually fragmentary, except in swampy areas where complete submergence may occur suddenly. Plants become covered and in time fossilisation occurs. Although plants may not be visible on a cliff face, evidence of their former presence is provided by red stains caused by the oxidation of iron sulphide compounds. These have crystallised in plant cells, thus effecting the petrification of plants.

The earliest plant fossils were found in the Fig tree shale of South Africa. Organisms were found in chert which were similar to bacteria and the Eobacterium isolatum, and the blue-green algae, Archaeospaeroides bartonensis.

These spheroidal cells resemble forms of the Chroococcales. Further analysis of the chert shows the presence of two hydrocarbons, phytane and pristane, which indicates metabolic activity involving chlorophyll and, therefore, photosynthesis.

Stromatolites, dome-shaped calcareous bodies, were photosynthetic and similar in structure to the stromatolites that form today at intertidal waters as a result of the action of mats of blue-green algae. Oscillatoria are filamentous structures which are similar to blue-green algae. They show a higher level of organisation than the spheroidal plants as they consist of a series of cells united in a strand or filament.

The green algae found in other deposits of chert provide evidence of algae containing nuclei, Eocaryota, as distinguished from Procaryota. Botanical evolution occurs at four levels.

Level 1. This occurred during the late Palaeozoic and Mesosoic eras, three billion years ago, when the first plants were algae. These were followed four hundred million years ago with algae, bacteria and fungi. These plants dominated the Earth's vegetation. There was a wide variety of non-vascular plants which produced oxygen, food for animals, colonised many ecological niches and had a variable morphology. The algae possessed a sufficient potential for change and evolved into organisms that could live on land. These were able to combat the drying effect of the atmosphere and the loss of contact, with a constant supply of water and nutrients.

Level 2. This occurred during the early Devonian and through the Carboniferous era. Lower vascular plants evolved in many forms. The tree ferns dominated the land vegetation and set the stage for the origin of the seed plants. The first seed plants were replaced by new ones and the succeeding ones were replaced by the Gymnosperms. Dominance by lower vascular plants lasted one hundred and fifteen million years.

Level 3. This extended from the Permian to the Cretaceous, a period of one hundred and eighty million years. The new and morphologically advanced forms were the Gymnosperms. However the algae, herbaceous lycopods, horsetails, ferns, fungi and bacteria were all abundant.

Level 4. This period extends into the present day with the domination of the Angiosperms.

Theories of evolution

The Linnaean natural system of classification implies a gradual change among living organisms.

Buffon suggested that the environment of an organism plays a part in forming a new organism.

Lamarckism suggested a law of use and disuse and the transmission of acquired characters. Lamarck stated that any structure, if used to excess, would become strengthed to meet the new conditions and that these developments would be handed on to the offspring. Similarly structures which are not used become less developed.

Darwinism puts forward a theory of natural selection a) a struggle for existence, b) the survival of the fittest and c) the offspring will inherit any advantage the parent had to survive, and in this way **new species** are formed.

Genetic units are called **demes**. All the different genes in a deme form a **gene pool**. During interbreeding there is a gene flow but the overall picture will remain the same as a genetic equilibrium occurs. Such an equilibrium will be maintained until a mutation or interbreeding occurs at the boundaries with members of another deme. **Genetic drift** will occur as a result. Once a change occurs natural selection enables evolution to proceed. Isolation will occur which will eventually give rise to a new species.

Practice Questions

1. Who was the founder of modern genetics?
2. What is monohybrid inheritance?
3. Write out your own example of a monohybrid inheritance.
4. Distinguish between phenotypes and genotypes.
5. What is the meaning of the term that is opposite to dominant?
6. Distinguish between homozygous and heterozygous.
7. What is a backcross?
8. Give your own example to show the meaning of the term 'test cross'.
9. Problem:- Mr. Lloyd crossed pink flowered antirrhinums together hoping to increase his stocks of pink flowers. When the seeds grew 25% produced red flowers, 25% white flowers and 50% produced pink flowers. a) What should he do next to make sure that all the seeds he sows will produce pink flowers? Explain your results in symbol form. b) Mrs. Lloyd wanted a flowerbed of white antirrhinums only. How could she be sure of obtaining suitable seeds?
10. What is dihybrid inheritance?
11. Write out your own example of dihybrid inheritance.
12. Problem:- Mr. Lloyd used the pollen from a true-breeding, smooth podded pea plant with yellow flowers to fertilise the flowers of a true-breeding, wrinkled podded pea with white flowers and planted the resulting seeds. Only smooth podded, yellow flowered plants grew. By using Punnett Squares show what progeny would result if this F_1 generation was a) selfed and b) test crossed.
13. Distinguish between chromosomes and genes.
14. Explain a) linked genes and b) lethal genes.
15. What do you understand by variation?
16. Give an account of the sources of variation.
17. Write an essay on evolution in botany.
18. What is a gene pool?
19. How do new species arise?
20. Write an essay on 'organic evolution'.

6 Plant community

This is a group of naturally occurring plants living in a common environment, interacting with each other through food relationships.

Ecosystem

An ecosystem is a community involving both the non-living environment and the biotic community. The biotic community dies, decomposes and the simple inorganic and organic substances become the food for the new plant growth. This forms the basis of **food chains** for the other organisms that live in the community.

Producers	→ consumers	→ decomposers
autotrophs	heterotrophs	heterotrophs
(plants)	(animals)	(bacteria and fungi)

Food webs indicate the paths of energy flow through a community. At every trophic level there is great energy wastage. At the herbivore level wastage is caused by the plant material not being completely consumed, digested or assimilated. Food that is assimilated is used for metabolic activities. The energy loss is so high at each trophic level that it causes food chains to be small.

Climatic factors

Rainfall, humidity, wind, temperature and light combine together to form a definite type of climate with its own vegetation. Rainfall percolates through the soil. Excessive rainfall results in leaching and excessive evaporation in flushing. Atmospheric humidity affects the rate at which water is lost from the surfaces of plants and animals. Water loss is low when humidity is high. Wind affects the rate of transpiration and evaporation from moist surfaces.

Most living organisms have an optimum temperature range. Low temperatures slow down the metabolic rate while higher temperatures could be lethal. Light is important for photosynthesis and provides food for more complex animal life. Light also affects an organism's behaviour. Some communities change over many years so that successive communities displace one another in turn. The history of a plant community is called a **plant succession**.

A primary succession begins with the colonisation of bare soil by pioneer populations. Plant successions are named according to the conditions in which they start.

A succession of communities produced on dry soil is a **xerosere**. A succession of communities which begins on soil in shallow water and is gradually raised above the water by plant debris is a **hydrosere**. Both seres produce a **climax community** whereas interference by man or beast e.g. through grazing produces a **biotic sere**.

Secondary succession starts when the vegetation has been destroyed and where soil is always present.

On sand dunes or salt marshes it is possible to see the stages of a primary xerosere or hydrosere e.g. from the newly colonised fore dunes to the biotic climax in the grasslands behind the dunes.

Edaphic factor

Soil

Soil is formed from the weathering of rocks by frost, wind, temperature fluctuations and the chemical and mechanical effects of water and living organisms.

Soil is a complex substance composed of a mixture of substances:

Inorganic material–	forms the bulk of the soil and consists of material derived from the parent rock, including mineral salts which have dissolved out of surrounding rocks
Organic material–	humus composed of dead decaying organic material and or living plant matter (see later)
Water–	adheres to the soil particles
Bacteria and soil–organisms–	the microscopic animals which increase the fertility of the soil and larger animal fauna e.g. the earthworm
Gas–	found in pore spaces in the soil and provides a medium for living organisms.

There are many different types of soil but they can be divided into heavy i.e. clay and light i.e. sand.

A comparison between a heavy and light soil

Heavy i.e. clay

1. Small particles
2. Small air spaces
3. Retains water
4. Slow drainage
5. High capillarity
6. Cold soil
7. Contains a variety of mineral salts
8. A closely packed soil
9. Micro-organisms present

Light i.e. sand

Large particles
Large air spaces
Does not retain water
Good drainage
Low capillarity
Warmer soil
Lacks mineral salts

A loosely packed soil
Soil too acid for the presence of micro-organisms

Humus
Humus is composed of dead and decaying organic material. The type of humus depends on the presence of micro-organisms. **Mor humus** is formed in acid conditions and the calcium content is low. The soil is too acid for animals. **Mull humus** is formed where the calcium content is high. The soil has often passed through the alimentary canal of animals e.g. earthworms. Mor and Mull humus give rise to well-drained soils.

Where drainage is poor aerobic organisms cannot live and so organic material remains undecomposed and is referred to as **litter**.

A thick layer of litter is called **peat**.

Marl
Marl is a soil composed of limestone and clay or limestone and sand.

Loam
This soil contains both clay and sand particles, thus combining the properties of both types of soil. The clay provides the mineral salts and the ability to retain water, while the sand particles lighten the texture of the soil making it more friable i.e. the soil crumbs improve aeration and prevent water logging.

An acid soil
Water logged soils tend to be acid. A lack of air prevents the activity of the aerobic bacteria. A lack of aerobic bacteria means that the organic content of the soil increases due to the lack of decomposition, thus deeper layers of soil become acid.

Fertiliser to a clay soil
Humus is added to heavy soils in the form of manure. This extra humus makes the soil light and easier to handle. It has a low density, the air spaces are large and it has great water-retaining powers.

Flocculation
The addition of lime to a clay soil makes the particles stick together to form larger particles.

Leaching
The soluble materials in a soil can be washed out of the soil with the draining water and so become out of the reach of the plant roots. This is called leaching and is responsible for reducing soil fertility. Leaching can be prevented by planting crops which absorb the mineral salts while growing and which can be ploughed in and left to decay. Any loss of minerals can be compensated for by the addition of fertilisers.

Soil colour
The colour of the soil is dependent on the humus content, the iron content and the presence of other minerals. When the humus is plentiful the colour is brown/black as the humus particles stick to the mineral matter and mask their effect. Soils of small humus content are much lighter in colour. Poor drainage and poor aeration, the presence of iron and anaerobic decomposition give the soil a grey colour. The improvement of this soil is brought about by oxidation and the iron content produces a red soil which is low in humus.

The colour of the soil affects the temperature of the soil. A dark soil absorbs more solar radiation than a light soil. The presence of humus in the soil also has a diluting effect because the higher water content makes the soil cold.

Experimental work on soil
1. To determine the properties of the solid components in a soil.
2. The percentage weight of solids in a soil.
3. The percentage volume of air in a soil.

4. Determine the percentage weight of humus in a soil.
5. To compare the permeability of sand, clay and loam.
6. To determine the capillarity of sand, clay and loam.
7. To demonstrate the presence of micro-organisms in the soil.

Pollution

Pollution is the addition to our environment of a substance or form of energy that is harmful to life, and at such a rate that the environment fails to absorb it. Water, air and land are all polluted in some way and often as a result of man's mismanagement.

Water pollution
1. Industrial pollution caused by industrial chemicals and wastes being deposited in nearby rivers or dumped at sea.
2. Sewage contamination.
3. Pesticides and herbicides washed into rivers from nearby fields.
4. Artificial fertilisers also washed into rivers from nearby fields.
5. The use of water for cooling power stations raises the temperature of the water into which it is discarded.
6. Tanker disasters lead to oil polluting the seas which are further contaminated by the detergents used to scatter the oil.

Water pollution leads to a change in the flora and fauna of the water and often the death of the whole ecosystem. Organisms are destroyed by poisonous chemicals, fluctuations in temperature or by lack of oxygen.

Land pollution
1. Industrial pollution caused by the indiscriminate dumping of chemical wastes.
2. Pesticides and herbicides used to control pests and weeds.
3. Land misuse caused by despoiling the countryside to obtain vital fuels and minerals.
4. The loss of land to buildings which often deface the countryside.
5. Household pollution caused by packaging and sewage disposal.

Land pollution destroys the flora and fauna of the land, interfering with the balance of nature.

Air pollution
1. Industrial pollution is caused by the emission of smoke, dust and gases. Dust causes damage to the respiratory surfaces, coating or irritating

the bronchioles and lungs, accelerating bronchitis or causing pneumoconiosis or asbestosis. Crops become coated with a fine dust, therefore decreasing their photosynthetic activity.

Smoke and dust prevent the sun's rays from penetrating through to the organisms below, thus preventing photosynthesis and preventing vitamin D production in man.

2. Household pollution caused by smoke from domestic fires and bonfires.
3. Exhaust pollution caused by the internal combustion engine. Vast quantities of carbon monoxide are produced. Carbon monoxide interferes with the haemoglobin content of the blood which may lead to anoxia and an increase in coronary heart disease. Lead is an accumulative poison. It interferes with enzymes which can lead to congenital side-effects in children.
4. Radioactive pollution may be caused by nuclear reactors or nuclear weapons. Abnormal doses of gamma radiation increase the rates of gene and chromosomal mutations. Radiation sickness, carcinoma and sterility are other effects.
5. Noise pollution caused by power tools, jet airliners and noisy radios may lead to impaired hearing.

Conservation

Conservation involves the planned use of land, water, air and all other natural resources so that they are not wasted or used foolishly.

Practice Questions

1. What is an ecosystem?
2. Give your own example of a food chain.
3. What do you understand by the term food web?
4. Why are food chains usually small?
5. Write an essay on plant succession.
6. What is soil?
7. Tabulate the differences between a clay soil and a sandy soil.
8. How does loam improve a soil?
9. Why is a water-logged soil acid?
10. Why is the colour of a soil a good indicator of the type of soil present?
11. What do you understand by the term pollution?
12. Write an essay on water pollution.
13. Enumerate the causes of land pollution.
14. Give an account of pollution caused by the combustion engine.
15. Write an account on any aspect of conservation that interests you.

7 Classification, Bacteria, Viruses, Fungi, Lichens and Algae

Classification

Taxonomy is the study of the classification of living organisms according to their similarities and differences. A scientific classification of living organisms is very precise and detailed involving morphology, anatomy, physiology, cytology, genetics and ecology of the organisms.

Linnaeus, a Swedish botanist in the eighteenth century, is responsible for our present-day system of classification. Latin names were used because Latin was the common language of all scholars. All living organisms are divided into large groups called the animal and plant kingdoms. A **kingdom** is therefore the highest taxonomic rank.

Plants are grouped or classified into a **division** (phylum) because of a common body plan. Each division is subdivided into **classes** because of variations in this common plan. Each class contains **orders**, each of which follows a certain way of life. Orders are subdivided into **families** which are subdivided according to **genus**. Each genus contains many closely-related **species**. Organisms that are grouped in a species have many characteristics in common and are therefore a group of individuals which are able to breed among themselves. A species is the smallest taxonomic rank.

Classification of the creeping buttercup

Kingdom	Plantae
Division	Spermatophyta
Class	Angiospermae
Order	Ranales
Family	Ranunculaceae
Genus	Ranunculus
Species	repens

A binomial nomenclature of system is a method of naming a particular species of plant. The generic name is written first followed by the specific name e.g. the creeping buttercup – Ranunculus repens, as compared to the meadow buttercup – Ranunculus acris. A natural classification is based on common resemblances. The more features organisms have in common the more closely are they related by descent. A natural system of classification may therefore indicate an evolutionary relationship.

Division Schizomyophyta (Bacteria)

General characteristics
Bacteria are small acellular or unicellular organisms of varied shapes. They can be spherical-coccus, rod-bacillus, curved-vibrio or spiral-spirilla. Some are parasitic and harmful while others are free-living and beneficial. Each bacterium is composed of a cell wall enclosing nuclear material. They reproduce by simple fission. Rapid reproduction occurs under favourable conditions.

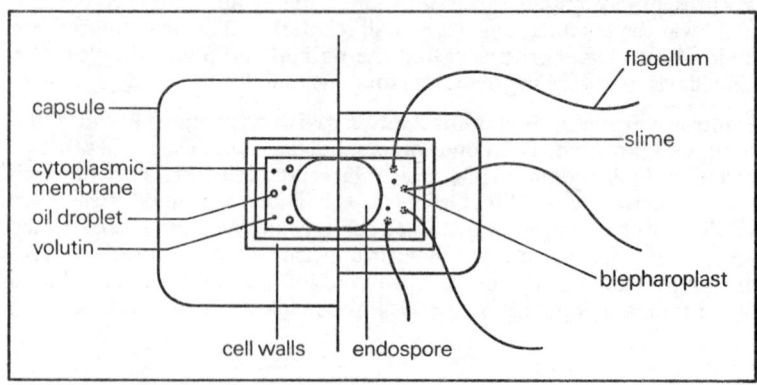

Bacteria in Nature
Bacteria are active agents in the decomposition of dead organic material which releases the elements needed for the growth of plants. They decompose the wastes of man and animals. Bacteria purify sewage, both aerobically and anaerobically, into harmless materials which can be used as fertilisers for the new growth of plants. Bacteria are involved in both carbon and nitrogen cycles.

Bacteria in dairy products
Milk is an ideal food for mammals. It contains few bacteria. Upon standing it becomes sour and the milk curdles. Lactose is changed into lactic acid by bacterial action. Cheese consists of the curd of milk which is separated from the liquid portion of the milk. The variety of cheese depends on the kind of organism used to ripen it, e.g. Swiss cheese contains large holes made by the gas produced by bacteria. One kind of

buttermilk is produced by growing Lactobacillus acidophilus. In the manufacture of butter large numbers of bacteria are carried with the fat globules and contribute to the flavour.

Cheese can be spoiled by bacteria e.g. Cheddar is discoloured by Lactobacillus rudensis, and Scropulariopsis brevicaulis grows on Camembert producing a strong ammoniacal flavour.

Bacteria in food
A common cause of food spoilage is enzyme action caused by the presence of bacteria. Damage may cause slight wastage, total loss or may lead to serious illness. To preserve food from the action of bacteria and thus avoid infection the food should be desiccated, refrigerated or treated with food preservatives.

Meat and fish spoilage is caused by the formation of bacterial slimes on the surface. Stored fruits and vegetables are susceptible to attacks by bacteria and will be lost if the bacteria is not checked. Food poisoning may be caused by Clostridium botulinum.

Bacteria and disease
Bacterial diseases of man and animals are more common than diseases caused by fungi. This reverses the condition found in the plant world. Pneumonia is caused by Diplococcus pneumoniae, cholera by Vibrio comma, and bubonic plague by Pasturella pestis.

Bacteria enter the host via the respiratory tract, the intestinal tract, by direct or indirect contact, abrasions of the skin and via insect punctures. The success or failure of the infection depends upon the virulence of the organism, the numbers of the invading bacteria and the susceptibility of the host.

A harmless bacterium, Escherichia colia, is an indicator bacterium. Its presence stimulates the search for the presence and activity of the other harmful bacteria.

Bacteria can be grown on nutrient agar in petri dishes.

Viruses
Viruses contain only DNA or RNA and are single macromolecules of nucleo-protein. Viruses are composed of a protein coat surrounding nucleic

acid. They are exceedingly small, neither cellular nor motile, and can be crystallised. Viruses will only reproduce within living cells.

Viruses are classified by the type of cells they infect; plant viruses which have a core of RNA, animal viruses which have a core of DNA or RNA and bacterial viruses or bacteriophages (phages). Viruses are obligate parasites and are best known for the diseases they cause.

The tobacco virus is a rod-shaped virus about 300 nm long. It is composed of a hollow protein coat surrounding a core of RNA. The presence of the tobacco mosaic virus was apparent when sap of an infected plant was injected into a healthy plant and caused the symptoms of the disease to appear. The leaves became green and mottled. On examination of the sap obtained from the plant by sucking insects, a virus containing a central helix of RNA was found to be the infective agent. Viruses are transferred from plant to plant by insects. Human virus diseases are cold sores, caused by contact with the virus Herpes simplex, and measles, caused by a virus spread by coughing and sneezing.

Domestic animals also suffer from viral diseases. Rabies is caused by a virus and can be transmitted to man from the bite of a rabid dog. Bacterial viruses are called phages. The phage has a six-sided head with a hollow tail. The nucleic acid is DNA. The tail is plugged to prevent the DNA escaping. The tail may have fibrils which attach themselves to a bacterium during infection.

Phage infection
Phages infect bacterial cells by becoming attached to them by the fibrils of the tail. The cell wall is penetrated and the DNA is injected into the bacterium, leaving the protein coat behind. The virus DNA takes over the vital functions of the bacterium. New and different enzymes are formed to produce virus DNA and protein coats. Many viruses are produced in a short period of time.

Division Eumycophyta (Fungi)

General characteristics
Fungi are mainly terrestrial. The body is called a mycelium and is composed of filaments called hyphae. The cell wall is composed of fungal cellulose containing chitin. There is no chlorophyll present. All are heterotrophic, either parasites or saprophytes.

Class Phycomycetes

Order Mucorales
e.g. Mucor mucedo – the pin mould fungus.
This is a saprophytic fungus living on dead or decaying organic material.

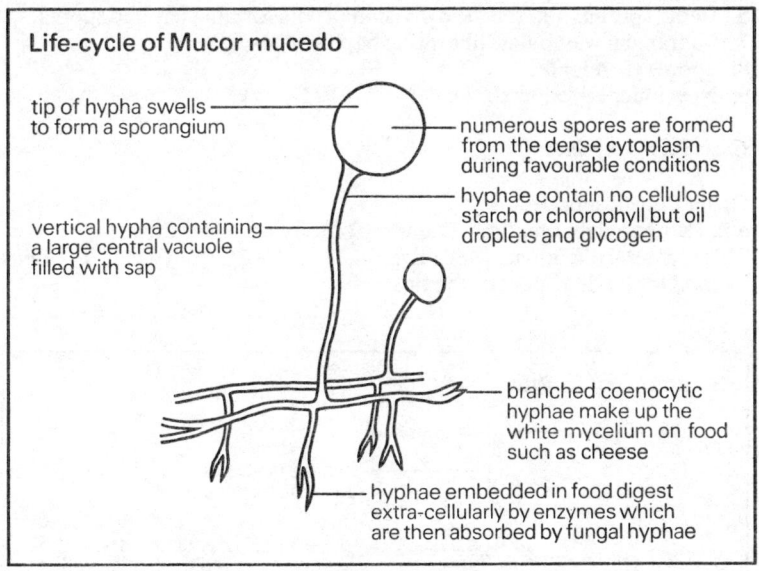

Asexual reproduction

1. **Columella** swells and the subsequent pressure helps to break the wall of the **sporangium.** The light spores are liberated and spread by the wind.
2. Haploid spores germinate in warm, damp conditions.
3. New mucor produced.

Sexual reproduction

1. Two short hyphae from different strains meet each other.
2. **Gametangia** are formed at end of each hypha. The intervening wall breaks down and the nuclei mix and fuse together forming diploid nuclei.

3. A **zygospore** forms covered with a thick, black, hard outer coat (presence of calcium oxalate).
4. The zygospore is liberated when the hyphae shrivel. It remains dormant until the return of favourable conditions.
5. The zygospore germinates liberating a single pro-mycelium. A **diploid** generation.
6. In the sporangium reaction division occurs forming haploid spores.
7. Sporangial wall bursts liberating haploid spores.
8. Spores germinate.
9. New mucor produced.

Class Phycomycetes

Order Peronosporales
e.g. Pythium debaryanum – Damping off disease of seedling.
This is a parasitic fungus which lives in or on a host, from which it obtains its food to the detriment of the host.

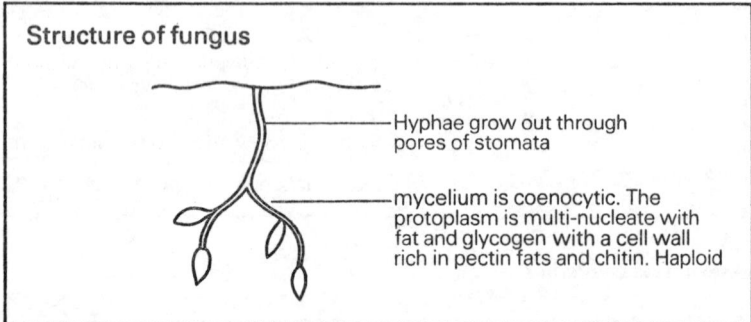

Asexual reproduction

1. Tips of *hyphae* constrict to form **sporangia,** each sporangium containing numerous *spores*.
2. Sporangia are easily detached from the fungus – dispersed by e.g. wind or raindrops.
3. **Under dry conditions:**
   Sporangia function as **air-borne spores** called **conidia. Haploid.**

4. **Under wet conditions**:
   Contents of each sporangium **divide** into several **bi-flagellate** kidney-shaped **zoospores** which are liberated from the sporangium and may swim freely in a film of water.
5. In the presence of moisture and a suitable host: **conidia** and **zoospores** may **germinate**. A fine tube or *hypha* is extended which penetrates the leaf cuticle and epidermal cell. Nutrients are absorbed from the plant cell and branched hyphae or haustoria grow between the plant cells, absorbing food from the host.
6. Before the host is killed **hyphae,** emerging through e.g. stomata on the surface of the leaf, produce fresh sporangia.

Sexual reproduction

1. **Sex organs** develop on the hyphae, occurring internally or externally.
2. **Male** branch swells to form a club-shaped **antheridium** while the globose **female oogonium** develops at the end of a hypha.
3. A short **conjugation tube** develops between the antheridium and the oogonium.
4. **Fertilisation** occurs when the antheridial contents pass into the oogonium to fuse with the female contents to form the **oospore.**
5. Oospore becomes coated with a thick **protective coat.**
6. When host dies and decays, oospore falls to the ground and passes into the soil.
7. In the presence of **moisture** and a **host,** the oospore **germinates**. It undergoes meiosis to form **zoospores** or put out **germ tubes.**
8. After germination the fungus grows throughout the plant.

Class Ascomycetes

Order Endomycetales
e.g. Saccharomyces cervisiae – Yeast

Sexual reproduction

This occurs when conditions are unfavourable. Each cell acts as a sporangium. The nucleus divides into 4 ascospores. Each ascospore is haploid from a diploid nucleus. They are resting spores and to regain their diploid nature two must fuse together either before or after they germinate.

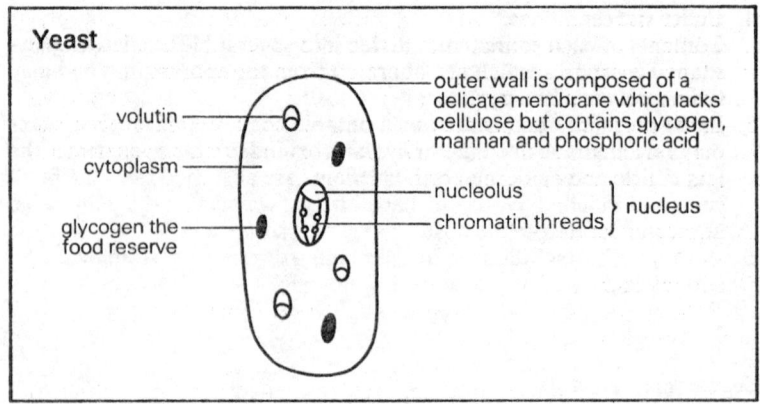

Asexual reproduction

This occurs when conditions are favourable. Yeast buds. An outgrowth of the parent gradually assumes the parent. Colonies of yeast may form if individuals are not separated from parent.

Class Ascomycetes

Order Pezizales
e.g. Pyronema

Pyronema is a saprophytic fungus which lives on burnt soil. The mycelium forms a white fluffy layer on the surface and develops male and female sex organs. The female consists of a round multinucleate base, the ascogonium, and an elongated, curved cell, the trichogyne, at its upper end. On an adjacent hypha a club-shaped antheridium arises. The antheridium comes into contact with the trichogyne, the nuclei disintegrate, the basal wall of trichogyne breaks down and the antheridial contents mix with the ascogonial contents. Multinucleate pairing of male and female nuclei occurs but no fusion.

After fertilisation the ascogonium becomes cut off from the trichogyne and branching ascogenous hyphae grow up from the new wall. These give rise to asci. Sterile hyphae mingle with the asci and become surrounded by a fleshy ascocarp. The ascocarp is disc-shaped, red/yellow and two to three millimetres in diameter. The asci and paraphyses form the hymenium and an open ascocarp. an apothecium.

The paired nuclei pass into the ascogenous hyphae where they multiply. The terminal cell of a branch that is to become an ascus bends back to form a hook and the nuclei divide. Three cells are cut off by walls. The terminal and basal cells are uninucleate but the middle one has become two nuclei of opposite sex. These fuse and the middle cell becomes the ascus. The fusion nucleus divides to form eight ascospores. The ascospores are liberated to form new mycelia on burnt ground.

Class Basidiomycetes

Order Agaricales
e.g. Psalliota campestris – the common field mushroom
This is a saprophytic fungus.

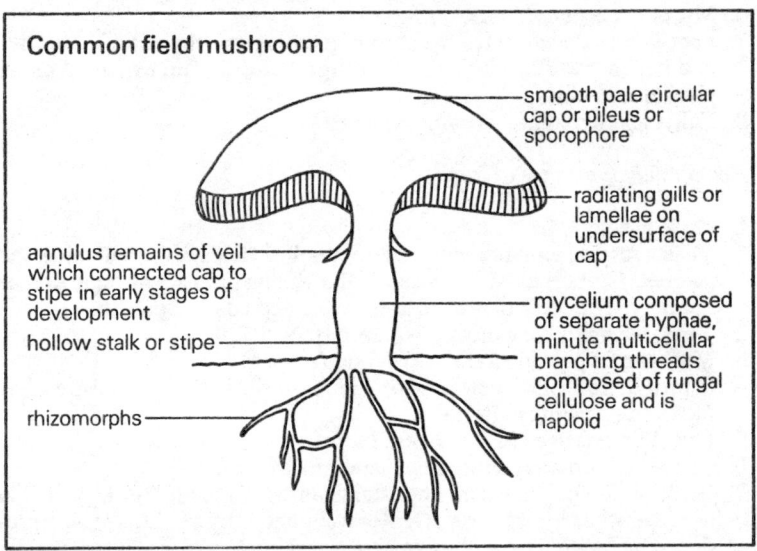

A gradual elaboration of asexual reproduction occus in the fungi.
1. The central part of each gill is composed of loosely-packed hyphae whereas the cells on the lateral surfaces are a smaller, bi-nucleate and more closely packed **sub-hymenial layer**.
2. The sub-hymenial layer produces sterile **paraphyses** or hairs mixed with club-shaped, bi-nucleate spore bearing **basidia.**

3. Spores are produced by **nuclear fusion**. Each basidium contains two nuclei which fuse. This forms a **diploid** nucleus. This nucleus undergoes **meiosis** to produce four **haploid nuclei** and four **sterigmata** occur at the free end of the basidium. One nucleus passes into each sterigmata which swells to form the **spore** which is constricted from the rest of the structure. At the base of each spore is a small **hilum**.
4. As the spores ripen the gills change colour. The nucleus in each spore divides into two. When ripe the spores are projected into the space between the vertical gills which are sufficiently apart to allow the spores to fall below the gills and be blown away by a gentle breeze. Before discharge each space exudes a drop of water from the hilum and this is carried away with the spore.
5. Many spores are produced and some will germinate to produce **mycelia**.
6. Mycelium is **homothallic**.
7. **Sporophore** develops from a mass of **pseudoparenchymatous** material and becomes differentiated into a **stipe** and **cap**. The hymenial tissue gives rise to gills.
8. Fully-developed mushroom produced.

Economic importance of fungi

Useful characteristics
1. Yeasts respire anaerobically. Fermentation yields carbon dioxide and alcohol. Carbon dioxide is used in the baking industry and the dry ice industry. The alcohol is used in the wine and beer industries.
2. Yeast is a valuable source of vitamin B.
3. Mushrooms provide a source of food for man.
4. Saprophytic fungi decompose organic material in the soil and are a key factor in recycling materials.
5. Penicillium is used in the cheese industry.
6. Citric acid is manufactured by Aspergillus niger.
7. Penicillium notatum is an important source of the antibiotic penicillin and the other antibiotics, streptomycin and tetramycin, are also produced by fungi.

Harmful characteristics
1. Rhizopus and penicillium spoil foods such as bread, cake and jam.
2. Fungi are pathogens of plants causing a loss of crops e.g. Puccinia graminis causes wheat rust. Claviceps purpurea causes ergot disease in rye.
3. Fungi attack timber e.g. Merulius lacrymans, the dry rot fungus.

Mucor mecedo damages leather, and other mildews attack fabrics and paper.

Sub division Lichenes

A lichen is composed of a fungus and an alga in close association. Each lichen has a specific fungal and algal component. In three genera the fungus is a basidiomycete and in others an ascomycete. The alga may be the cyanophyta or chlorophyta, usually non-filamentous. Both alga and fungus benefit from the relationship, a symbiotic relationship. The fungus absorbs and retains moisture while the alga is photosynthetic.

Internally Lichens may be divided into four regions, an upper region in which fungal mycelia are closely interwoven and between is gelatinous material which holds the fungi together. Below this is the algal region or gonidial region. Then comes the medulla in which the hyphae are loosely interwoven and then bounded by compact hyphae or the lower cortex.

Reproduction is 1) vegetatively by soredia which consist of a few algal cells surrounded by fungal filaments, which are easily removed from lichen surfaces, 2) sexual reproduction which occurs in the fungal partner only. Asci produce eight ascospores enclosed by an apothecium. The ascospores germinate but must make contact with their algal partners to survive.

Lichens are the first colonisers of rocky places. They bring about corrosion. On the death of the lichens the remains become the pioneer organic material. They are a source of food for man and animals and provide the raw materials for the manufacture of litmus.

Algae

General characteristics
Algae are mainly aquatic. The plant body is differentiated into stem, root and leaves and is called a thallus. The thallus may be unicellular e.g. Chlamydomonas, or multicellular with a flat dichotomously branched body e.g. Fucus. There is differentiation of the internal structure in Fucus compared with the anatomy of Spirogyra. There is a reduction in asexual reproduction and an elaboration of sexual reproduction, e.g. isogamous gametes in Chlamydomonas and Spirogyra, to anisogamous gametes or oogamy in Vaucheria and Fucus.

Classification of Algae

Division Chlorophyta   e.g. Chlamydomonas, Ulva
Division Phaeophyta   e.g. Brown algae e.g. Fucus
Division Rhodophyta   e.g. Red algae
Division Bacillariophyta   e.g. Diatoms
Division Cyanophyta   e.g. Blue-green algae

Division Chlorophyta
Order Volvocales
e.g. Chlamydomonas

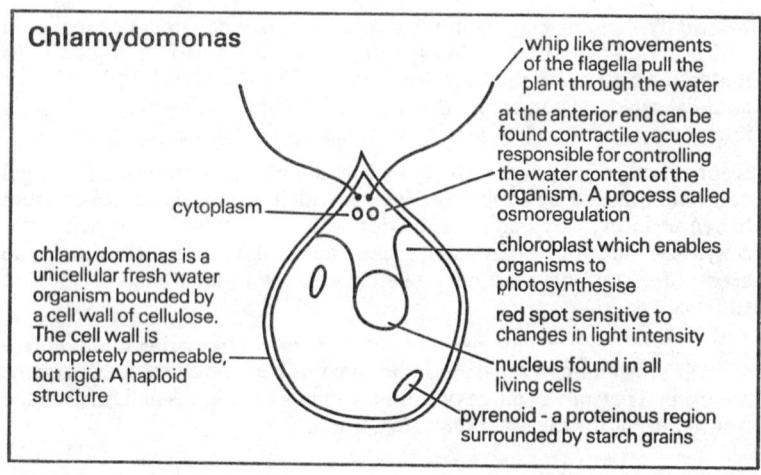

Life cycle of Chlamydomonas

Asexual reproduction
1. At maturity and during favourable conditions the chlamydomonas absorbs its **flagella** into the **cytoplasm**. The cell divides longitudinally and then divides into 2, 4 or 8 new individuals. A new cell wall forms around each one and gradually new chlamydomonas cells can be seen. The **parental wall** gelatinises and the new cells swim freely.
2. New chlamydomonas develops.

Sexual reproduction
1. At maturity and during unfavourable conditions chlamydomonas absorbs its flagella into the cytoplasm. The cell divides into 8, 16 or 32 bi-flagellate gametes which are all the same as the parent but have no cell wall.
2. The gametes are released into the surrounding water and each seeks out a partner from another parent.
3. **Conjugation** takes place. Two cells merge at their anterior ends. A **quadriflagellate zygote** remains motile for a few days before encysting. A **dyploid** structure.
4. A zygospore is formed which divides **meiotically** into four **bi-flagellate chlamydomonae**.
5. New chlamydomonas develops.

Division Phaeophyta
Order Fucales
e.g. Fucus vesiculosus – Bladder wrack

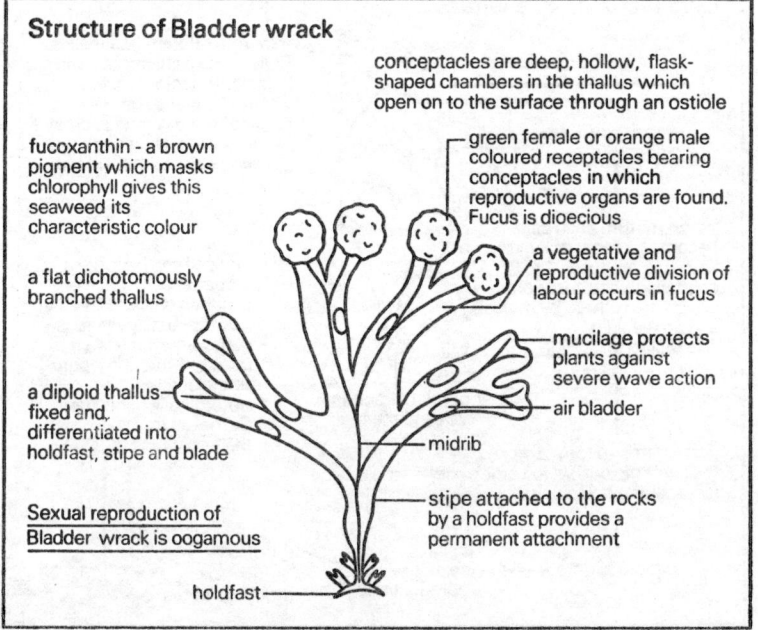

Structure of Bladder wrack

conceptacles are deep, hollow, flask-shaped chambers in the thallus which open on to the surface through an ostiole

fucoxanthin - a brown pigment which masks chlorophyll gives this seaweed its characteristic colour

green female or orange male coloured receptacles bearing conceptacles in which reproductive organs are found. Fucus is dioecious

a flat dichotomously branched thallus

a vegetative and reproductive division of labour occurs in fucus

a diploid thallus fixed and differentiated into holdfast, stipe and blade

mucilage protects plants against severe wave action

air bladder

midrib

Sexual reproduction of Bladder wrack is oogamous

stipe attached to the rocks by a holdfast provides a permanent attachment

holdfast

Division Chlorophyta
Order Ulotrichales
e.g. Ulva lactuca – sea lettuce

Ulva demonstrates a clearly defined alternation of generations in which a haploid, sexually reproducing thallus, the gametophyte generation, gives rise to a diploid, asexually reproducing thallus, the sporophyte generation. In Ulva the generations are identical and exhibit isomorphic or homomorphic alternation of generations. Reduction of the gametophyte generation is the general trend in the life histories of all higher plants, culminating in the Angiosperms where reduction is complete.

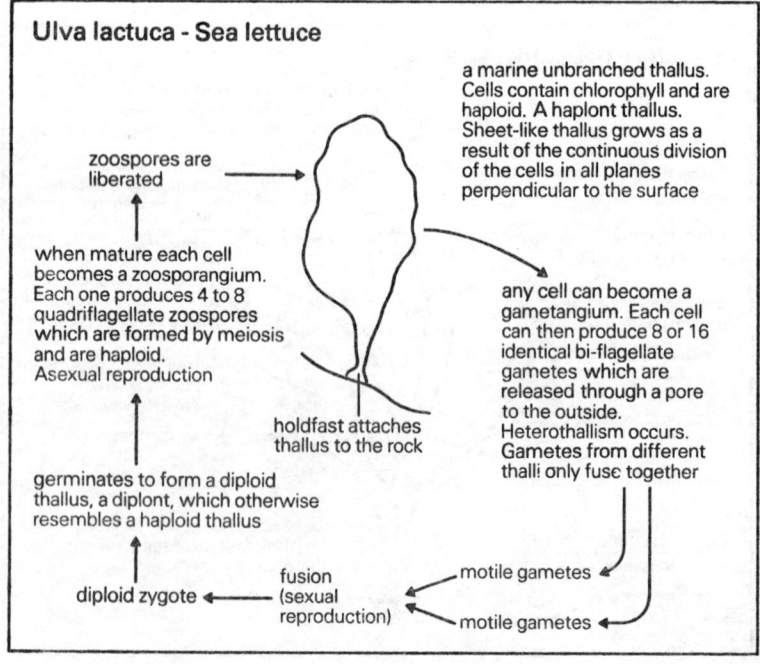

Life cycle

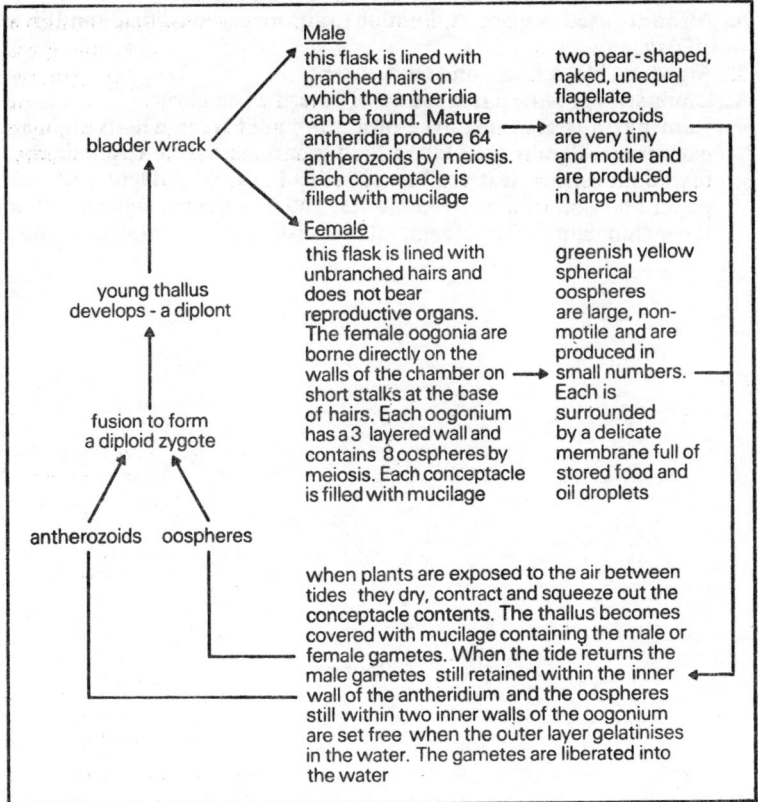

Economic importance of the algae
1. Algae are the basic food of food chains in water. The productivity of the sea depends on the phytoplankton.
2. Alginates are extracted from the Phaeophyta, forming gels used in the food and cosmetic industries.
3. Agar are produced from some Rhodophyta.

4. Seaweeds are used as fertilisers, fodder for animals and food for human consumption.
5. Ancient deposits produce our present day oil and gas.
6. Algae are used in space exploration to absorb carbon dioxide and give off oxygen.
7. Algae can block filters and taint water.
8. Laminarias provide a rich source of potash when burnt.
9. Laminaria is also a source of alginic acid from which algin is extracted. Algin is used in the food, pharmaceutical, cosmetic and textile industries. It is used as a coating in the manufacture of wax paper and non-inflammable fabrics, and as an emulsifying agent in the manufacture of ice-cream, salad dressing, soups and skin creams.

Practice Questions

1. What is the basis for a scientific classification of living organisms?
2. Name the major divisions in the plant kingdom.
3. What do you understand by a binomial system of classification?
4. Give a full classification of any species of plant you have studied.
5. Give the three main groups that are classified as Thallophytes.
6. Write an essay on the economic importance of bacteria.
7. What are the main characteristics of viruses?
8. Draw a generalised diagram of a bacteriophage.
9. Give an account of vegetative reproduction in Algae.
10. What do you understand by the term division of labour?
11. Distinguish with suitable examples between isogamy, anisogamy and oogamy in Algae.
12. What do you understand by an isomorphic alternation of generations?
13. Give an account of nutrition in Mucor.
14. Explain the terms a) coenocytic, b) hyphae, c) sporangium and d) haploid spores.
15. Distinguish between a saprophyte and a parasite.
16. Give an account of reproduction in any Basidiomycete you have studied.
17. Give an account of the economic importance of Fungi.
18. Tabulate the differences between Algae and Fungi.
19. What two plants form the Sub-division Lichenes?
20. How does sexual reproduction take place in Lichens?

8 Bryophyta

General characteristics

The green thallus is attached to the ground by rhizoids. The thallus is dichotomously branched and there is some differentiation of tissues e.g. for photosynthesis. Sexual reproduction takes place and there is heteromorphic alternation of generations. The vegetative plant is a haploid gametophyte with a sporogonium partially or wholly dependent on the gametophyte, the sporophyte generation. Bryophytes are chiefly terrestrial, making the first steps to land colonisation.

Anatomy of the gametophyte

Liverwort
The plant body is composed of haploid polyhedral cells. They are closely packed together. The upper and lower cortex contain many chloroplasts while the middle have fewer. The epidermis is composed of smaller cells.

Moss
The epidermis is composed of small, thick-walled cells enclosing a cortex of larger, isodiametric cells which contain chloroplasts. In the centre of the cortex there is a strand of smaller cells which are elongated longitudinally. This forms a simple conducting and mechanical tissue.

Division Bryophyta

Class Hepaticae
Order Jungermanniales
Sub-order Anacrogyne
e.g. Pellia epiphylla – a liverwort

Growth of the thallus is by the division of a single apical cell in a groove at the tip. The cell is protected from desiccation by mucilage-producing cells. New cells are cut off from three surfaces and the longitudinal division of the apical cell results in dichotomy.

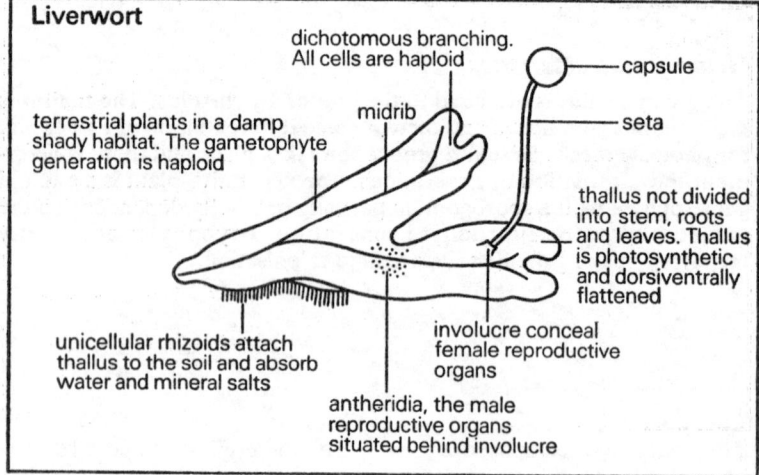

Sexual reproduction in Pellia

Female organs are archegonia which occur behind the thallus tip and are protected by an involucre. Each archegonium is composed of an outer wall, enclosing a developing oosphere contained in a round venter. Leading from the venter is a neck composed of a ventral canal cell and neck canal cells. At maturity the neck canal cell and ventral canal cells disappear and are replaced by mucilage. The oosphere is clearly seen in the venter.

Fertilisation

Fertilisation results in the formation of a diploid zygote. One zygote occurs beneath each involucre. The zygote develops into a diploid sporogonium which appears as an outgrowth of the vegetative thallus.

Male organs are short, stalked antheridia embedded in the tissues of the thallus and open to the exterior through a small hole. Each antheridium is an outer covering of cells protecting a mass of antherozoid mother cells. These divide meiotically to produce naked antherozoids. Each antherozoid has a spiral nucleus surrounded by cytoplasm, with two flagellae. When ripe the antheridium absorbs water, disrupts and then sets free the motile gametes. A motile gamete seeks out passive female oosphere which secretes a protein from the neck canal cells, attracting the male gamete. This is a chemotactic response.

BRYOPHYTA

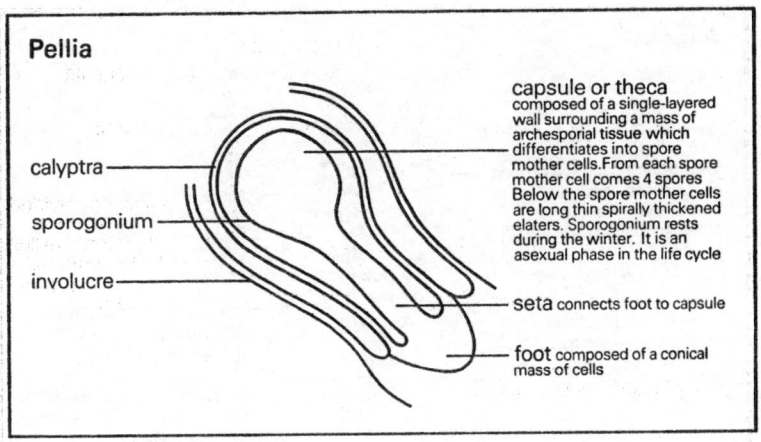

Pellia

- **calyptra**
- **sporogonium**
- **involucre**
- **capsule or theca** composed of a single-layered wall surrounding a mass of archesporial tissue which differentiates into spore mother cells. From each spore mother cell comes 4 spores. Below the spore mother cells are long thin spirally thickened elaters. Sporogonium rests during the winter. It is an asexual phase in the life cycle
- **seta** connects foot to capsule
- **foot** composed of a conical mass of cells

The sporogonium rests during the winter, an asexual phase of the life-cycle of Pellia. It is purely dependant on the vegetative thallus for its nutrition. The spring brings further development. The outer wall of the capsule dies. The seta elongates and ruptures the calyptra. Each spore within the capsule becomes a group of a few green cells and each may resemble a thallus. The spores are liberated by the capsule bursting into four as the wall dries out. As the capsule opens the elaters and spores are exposed. The elaters are hygroscopic and their movements loosen the spores which are dispersed by the wind.

Each spore germinates in damp conditions to produce a new haploid thallus, the gametophyte generation. Although terrestrial it is dependant on water for its reproduction and dry conditions for the dispersal of spores. Pellia exhibits a heteromorphic alteration of generations.

Class Musci
Order Bryales
e.g. Funaria hygrometrica – a moss

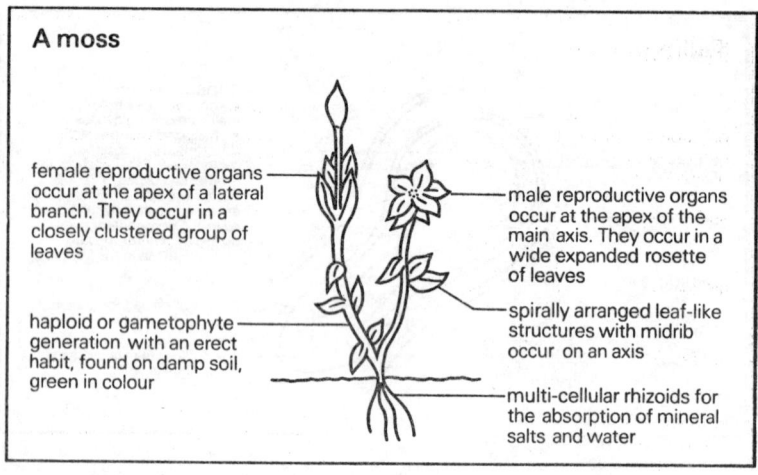

Sexual reproduction
The lateral apex shows **archegonia**. These are similar to Pellia with a stronger stalk and longer, twisted necks. They are interspersed with paraphyses. The oosphere can be clearly seen in the venter when the neck is open.
Fertilisation
Fertilisation results in a diploid oospore which develops into a sporogonium.

The male **antheridia** are clubshaped structures which occur on the main axis. They are interspersed with paraphyses which aid in water retention. Antherozoids develop similarly to the liverworts. The motile antherozoids seek out a female oosphere by chemotaxis.

When the sporogonium is mature and dry the seta is twisted. On wetting it unwinds quickly carrying the capsule with it. On drying it becomes coiled. When liberated the spores begin to germinate. The exosporium ruptures and the spore develops two or three delicate germ tubes. A branched filamentous structure called a primary protonema forms. After a period of growth some cells undergo rapid division to form a mass of cells called 'buds'. A new moss plant grows from these buds.

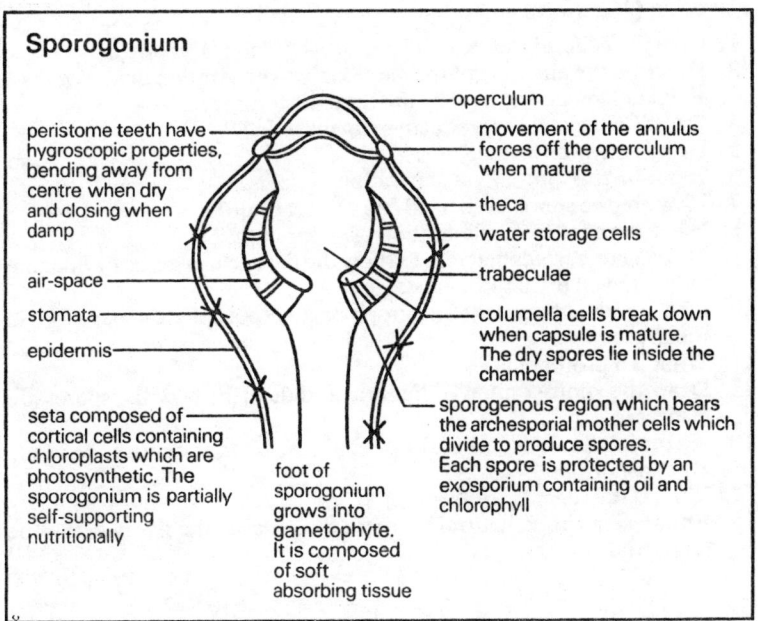

The gametophyte

The gametophyte reaches a climax in the liverworts and mosses where it partly complies with land requirements:-

1. It has a primary conducting system.
2. It remains small because of the absence of an internal aerating system.
3. Photosynthetic leaves are present.
4. It lives in a partially aquatic habit and as it has no cutin it must survive any unfavourable conditions in a dormant state.
5. It retains the ovum in the archegonium with access to a motile sperm.
6. Water is essential for fertilisation.

Practice Questions

1. Give the general characteristics of the Division Bryophyta.
2. Describe the anatomy of the gametophyte generation in Musci.
3. What do you understand by the term dichotomy?
4. Describe the male reproductive organs in Pellia.
5. How is a zygote formed?
6. Describe the sporophyte generation in Pellia.
7. How are the spores dispersed from the capsule?
8. What is meant by an alternation of generations?
9. Make a large, clear diagram to show the life cycle of Pellia or Funaria, indicating all the important stages.
10. Distinguish between the gametophyte and sporophyte generations in Funaria.
11. What is a protonema?
12. Draw the sporogonium of Funaria and describe how the spores are dispersed.
13. Explain the following a) operculum, b) trabeculae, c) seta, d) exosporium.
14. What is the function of the 'foot'?
15. Enumerate the structural features that enable the liverworts to be terrestrial.

9 Pteridophyta

General characteristics

These plants have roots, stems and leaves. Dichotomous branching may occur. There is internal differentiation and sexual reproduction occurs with antheridia and archegonia. Alternation of generations is evident.

Division Pteridophyta

Sub division Pteropsida
Order Filicales
e.g. Dryopteris felix-mas, the fern
The sporophyte fern plant is composed of a root, stem and leaves. There is a distinct division of labour. Ferns are homosporous but with no strobili. The terrestrial habit in plants has led to a variety of cell types. 1) mechanical tissue to support the plant 2) vascular tissue for the conduction of food and water 3) protective tissue to prevent desiccation.

Anatomy of the fern

Rhizome

The outer epidermis encloses an outer cortex of sclerenchymatous cells and an inner cortex of large, thin-walled parenchyma cells with starch grains. A cortex surrounds a central stele which encloses a pith. The stele has an intricate meshwork and is a dictyostele. Each small bundle is a meristele. Each meristele has a central core of xylem composed of scalariform tracheids with xylem paranchyma with dense, thin walls. The xylem is surrounded by phloem, a pericycle and endodermis. The tracheids are large, empty and polygonal. No xylem vessels are present. The elongated phloem cells are composed of sieve elements, without companion cells but with phloem parenchyma.

Root

The cells of the outer cortex are dead and enclose large, thin-walled parenchymatous cells which store starch grains. A central protostele is composed of xylem tracheids with two protoxylem groups in a diarch arrangement. Xylem has phloem on either side. Stele is surrounded by a double-layered pericycle and a single-layered endodermis. Lateral branches arise endogenously from the endodermis and not from the pericycle as in Angiosperms. There is no secondary growth.

Growth is by means of a single apical cell in the rhizome and root. In the stem apex there is a three-sided pyramidal cell with the base uppermost. New cells are cut from each of the surfaces which divide further before differentiation.

Sub-division Psilopsida
Order Equisetales
e.g. Equisetum arvense – Common horse tail

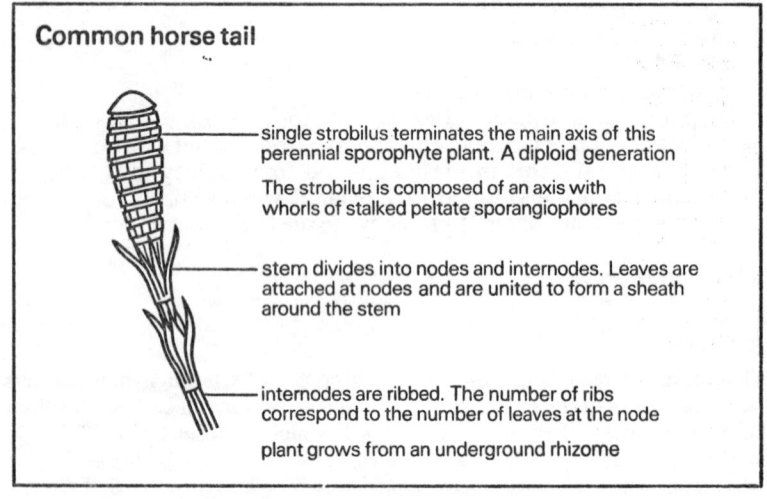

Common horse tail

- single strobilus terminates the main axis of this perennial sporophyte plant. A diploid generation
- The strobilus is composed of an axis with whorls of stalked peltate sporangiophores
- stem divides into nodes and internodes. Leaves are attached at nodes and are united to form a sheath around the stem
- internodes are ribbed. The number of ribs correspond to the number of leaves at the node
- plant grows from an underground rhizome

Each sporangiophore is umbrella-like in shape with pendant sporangia attached to a polygonal disc-shaped shield. The flattened lips fit closely together to protect the developing sporangia along with an annulus. At maturity the strobilus and sporangiophores shrink and the sporangia are opened. The outer wall of each spore splits into strips as it ripens and forms four hygroscopic elaters. These elaters are spirally coiled around the spore when dry and uncoiled when wet. The spores are homosporous. When they fall to the ground they germinate to form two types of prothalli, an archegonial prothallus and an antheridial prothallus. In

Life-cycle of the fern

The sporophyte fern plant is composed of a root, stem and leaf. A distinct division of labour occurs. It is homosporous and has no strobili

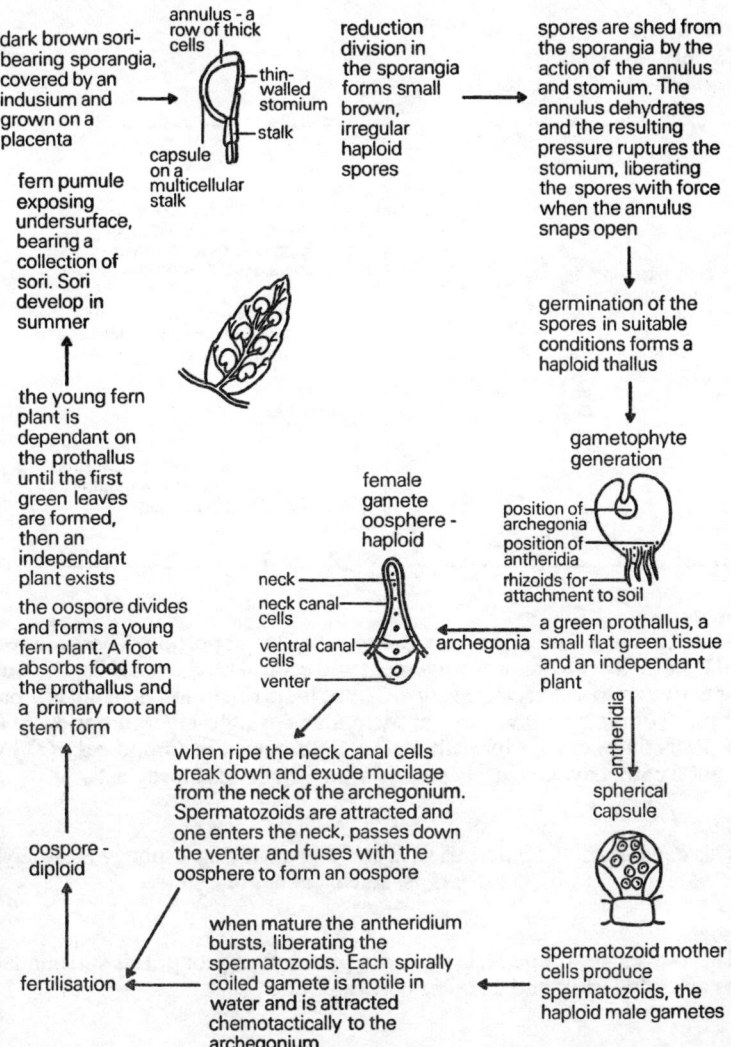

dark brown sori-bearing sporangia, covered by an indusium and grown on a placenta

annulus - a row of thick cells
thin-walled stomium
stalk
capsule on a multicellular stalk

reduction division in the sporangia forms small brown, irregular haploid spores

spores are shed from the sporangia by the action of the annulus and stomium. The annulus dehydrates and the resulting pressure ruptures the stomium, liberating the spores with force when the annulus snaps open

fern pumule exposing undersurface, bearing a collection of sori. Sori develop in summer

the young fern plant is dependant on the prothallus until the first green leaves are formed, then an independant plant exists

the oospore divides and forms a young fern plant. A foot absorbs food from the prothallus and a primary root and stem form

germination of the spores in suitable conditions forms a haploid thallus

gametophyte generation

position of archegonia
position of antheridia
rhizoids for attachment to soil

female gamete oosphere - haploid
neck
neck canal cells
ventral canal cells
venter

archegonia — a green prothallus, a small flat green tissue and an independant plant

when ripe the neck canal cells break down and exude mucilage from the neck of the archegonium. Spermatozoids are attracted and one enters the neck, passes down the venter and fuses with the oosphere to form an oospore

antheridia

spherical capsule

oospore - diploid

fertilisation

when mature the antheridium bursts, liberating the spermatozoids. Each spirally coiled gamete is motile in water and is attracted chemotactically to the archegonium

spermatozoid mother cells produce spermatozoids, the haploid male gametes

Equisetum two different gametophyte generations are produced. The male is slightly smaller than the female gametophyte. They are independent plants and resemble closely the development of Dryopteris felix-mas.

Sub-division — Lycopsida
Order – Selaginellales
e.g. Selaginella Kraussiana – a club moss

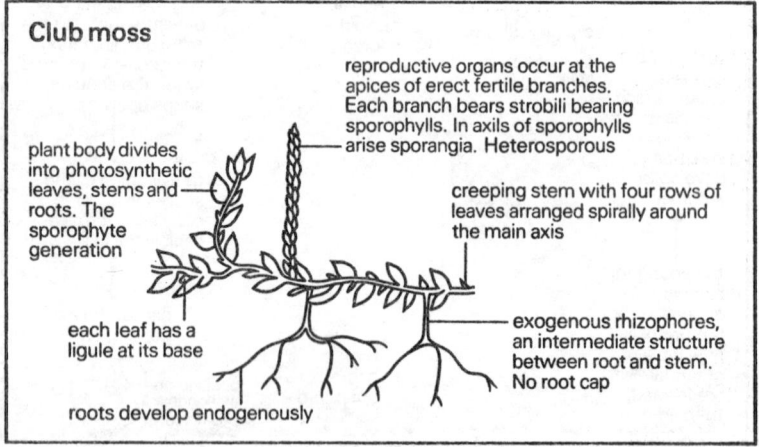

Stem
The stem has a flattened upper and rounded lower portion. There are two steles with a protostelic arrangement and central tracheidal cells with one or two exarch protoxylem groups. The phloem rings the xylem but is separated from it by a layer of parenchyma. A single-layered pericycle is linked to the cortex by trabeculae. The cortex is composed of thin-walled parenchyma, closely packed with a chain of chloroplasts.

Leaf
This consists of an epidermis with stomata enclosing a spongy mesophyll of parenchyma with chloroplasts. Each leaf has one stele.

Root
This is a single, round stele with one protoxylem group. It is surrounded by an endodermis and an outer cortical layer.

Growth
The main stem grows from two four-sided apical cells. Growth of a branch is from a single three-sided cell, formed from a portion of one of the main stem apical cells.

Selaginella is heterosporous. Microsporangia occur at the top of the plant and megasporangia occur at the bottom. Each sporangium arises from a group of meristematic cells at the stem apex and develops into a capsule borne on a short stalk. Each capsule has two layers of cells; the outer layer forms a sporangial wall and a row of larger cells, while the archesporium gives rise to sporogenous cells surrounded by tapetal cells. The sporogenous cells give rise to the spore mother cells.

Microsporangia
Spore mother cells separate and float free in a nutritive fluid formed by the disorganisation of the tapetal cells. Each gives rise to four microspores.

Megasporangia
Only one spore mother cell increases in size at the expense of the others. The spore mother cell produces four megaspores while the others have a nutritional function with the tapetal cells.

Dehiscence of the sporangia takes place on drying and the spores are liberated.

Germination of the spores
Microspores start germinating before liberation. The microspore increases in size and a small cell or male prothallus is cut off at the pointed end. The rest of the cell divides into eight peripheral cells which form the antheridium, and this will eventually have a nutritive function for the spores, surrounding two or four central cells. The central cells divide further to form the mother cells of the antherozoids. The gametophyte is not independent. A bi-ciliate antherozoid is formed which is haploid.

Megaspores start germinating before liberation. The nucleus divides into two. One passes to the apex of the spore and the other to the base. The base forms cells after germination, providing large cells full of food. The megaspore ruptures at the apex, exposing the short necked archegonium. It forms a small green female prothallus but is not independent. It is haploid.

Fertilisation
Fertilisation occurs when a motile antherozoid enters the ovum and fusion takes place. The resulting zygote invests itself with a wall and becomes an oospore, contained in the venter of the archegonium. The oospore divides to form the embryo sporophyte. An upper and lower cell is formed. The upper cell remains unicellular and forms a suspensor which pushes the developing embryo down into the tissue of the prothallus. The basal cell divides further into eight cells to form epibasal and hypobasal cells. From the four apical cells a stem apex and two cotyledons are formed, while the four cells of the hypobase give rise to a hypocotyl. The hypocotyl becomes enlarged into a 'foot' which has an absorptive function. The first root is adventitious, developed from the hypocotyl close to the suspensor. The stem and cotyledons leave the spore and grow above the ground into a sporophyte generation, the diploid.

Gametophyte in the Pteridophyta
There is a decline of the gametophyte and a rise in the sporophyte. The plants leave the water and approach land. They therefore need:-

1. Mechanical support for land habitation
2. Protection to prevent desiccation
3. An internal aerating system
4. Photosynthetic tissue
5. The ovum must be retained within the parent

The sporophyte develops as an independent plant. Heterospory evolves and the megaspore and prothallus are retained within the megasporangium, thus enabling the plants to live on land.

Practice Questions

1. Name three sub-divisions of the Pteridophyta and write the full names of each typical representative of each sub-division.
2. How does growth occur in the fern?
3. Describe the sporogonium of the fern.
4. Explain 'the gametophyte generation is independent'.
5. What do you understand by an alternation of generations?
6. Describe the process of fertilisation in the fern.
7. Explain the term homosporous and heterosporous.
8. Distinguish between antheridia and archegonia.
9. Make a large labelled diagram of Selaginella.
10. Describe the germination of the microspore in Selaginella.
11. Describe the development of the zygote in Selaginella.
12. What is the function of the 'foot' in Selaginella?
13. What evolutionary changes occur in the Pteridophyta?
14. Make a labelled diagram to show the life cycle of Selaginella, indicating the important stages.
15. Explain a) strobilus, b) sori, c) prothallus and d) elaters.

10 Spermatophyta

General characteristics

These are terrestrial plants which are divided into stem, root and leaves. All show internal differentiation. There is a definite alternation of generations. Sexual reproduction occurs with the reproductive organs borne on either strobili or flowers. Fertilisation is by a pollen tube (siphonogamous). The male gametophyte is shed as pollen grains and is transported by various agents to the female gametophytes. Both gametophytes are small. Fertilisation results in a seed which may be naked or enclosed but which is developed by the sporophyte to form a fruit.

Division Spermatophyta – seed bearing plants

Class Gymnospermae — produce naked seeds
Order Cycadales
e.g. Cycas revoluta

Cycas is a tropical plant and is considered to be a living fossil. It is composed of an unbranched, short stem with a cluster of leaves at its apex. Leaves are of two kinds. Some are very large with a central rachis bearing many pinnae and smaller leaves are dry and scaly bearing brown hairs. These leaves alternate in zones on the stem. The scaly leaves protect the leaves of the new foliage. The root is a short primary tap root equal in girth to the stem, with many lateral roots. Root tissue contains blue-green algal cells and bacteria and fungi occur in root nodules. Cycas is dioecious. Strobili of some sporophylls develop male microsporangia and others develop female ovules.

Male cone
Many woody, wedge-shaped microsporophylls are spirally arranged on an axis. The lower surface is covered with sori of microsporangia, each composed of a thick outer wall enclosing a tapetum and archesporial tissue. Each spore mother cell produces four microspores or pollen grains. Each small grain has a double wall, an exine and an intine. Spore

Female cone
Megasporophylls are spirally arranged in a group at the stem apex. Each megasporophyll resembles a leaf. The distal end is flat and pinnate whilst its proximal end bears large, naked ovules in two lateral rows. The nucellus caught by a pollen drop from the micropyle. The archesporial tissue of the nucellus forms four megaspores, of which three abort

output is great and spores are wind dispersed. The male gametophyte begins to develop before dispersal.
becomes surrounded by a soft integument which hardens. A beak-like micropyle allows access into the ovule. The nucellus forms a nucellar beak with a pollen chamber at its apex. The chamber collects microspores which are

Development of the male gametophyte

A prothallial cell is cut off inside the pollen grain and this is the vegetative structure. The remaining cell forms the antheridial cell and a tube cell with a prominent nucleus. This is when pollination occurs. The antheridial cell forms a stalk and a body cell or antheridium. The microspore wall bursts and the tube cell grows out. A pollen tube penetrates the nucellus and breaks down and digests its tissues. When the ovule matures the antheridium enlarges with blepharoplasts on either side of the nucleus. The body cell divides and each new cell and blepharoplast forms an antherozoid.

and the lower one forms the megaspore and enlarges to form the female gametophyte.

Development of female gametophyte

The megaspore enlarges at the expense of the nucellus and develops a thick, suberised membrane. Nuclear division occurs to form a tissue rich in starch. The integument differentiates into three layers, inner, outer and middle. On the female prothallus near the micropyle, archegonial cells arise which form the archegonia. Each is composed of a two-celled neck, a neck cell and an oosphere in the venter. A nutritive jacket forms around the oosphere from the prothallial cells. The oosphere cell divides to form a ventral canal cell and an oosphere whose nucleus enlarges. It is ready for fertilisation and the prothallial tissue around the archegonia grows upwards to form a ridge, so that the protruding neck lies in a shallow archegonial chamber. The megaspore membrane ruptures and the pollen tube hangs down into the chamber. The tube bursts, liberating the antherozoids with a fluid. When an antherozoid reaches the neck of an archegonia it is sucked down to the oosphere. Fusion occurs to form a zygote.

After fertilisation the zygote divides rapidly to produce many free nuclei within the oospore. Near the base of the oospore some of the cells elongate to push cells into the prothallial tissue. The embryo develops into a straight axis with the root towards the micropyle and with two cotyledons at the plumule end. The embryo is embedded in the food-storing remains of the female prothallus. The suspensors coil and twist to form a protective pad over the radicle tip. A seed coat is formed out of the two outer layers of the integument. Seeds are shed when ripe and germination is hypogeal.

Class Gymnospermae
Order Coniferales
e.g. Pinus sylvestris – Scots pine

Anatomy in Pinus
This is similar to the Angiosperms.

The epidermis of the stem covers a lignified hypodermis. The cortex is composed of parenchymatous cells. A single ring of vascular bundles occurs separated by medullary rays. These connect the cortex to the central pith. There is no endodermis or pericycle but a series of resin canals are present. Each vascular bundle is collateral i.e. xylem – cambium – phloem radially arranged. Xylem is composed of spirally thickened tracheids in the protoxylem and reticulately thickened tracheids in the metaxylem. Bordered pits occur in xylem vessels. The cambium separates xylem from phloem and is meristematic. Phloem is composed of sieve elements but no companion cells. The epidermis is replaced by a bark formed by a phellogen. Secondary tissue is formed.

Leaf
A cuticularised epidermis occurs with sunken stomata. A hypodermis of sclerenchymatous cells is formed beneath the epidermis. Parenchyma cells form the cortex and resin canals are present. There is a centrally situated stele surrounded by an endodermis and pericycle. These enclose two vascular bundles.

Root
This is similar to a dicotyledonous angiosperm. The steles are diarch or triarch. A root cap and root hairs are present. In young roots ectotrophic mycorrhiza occur.

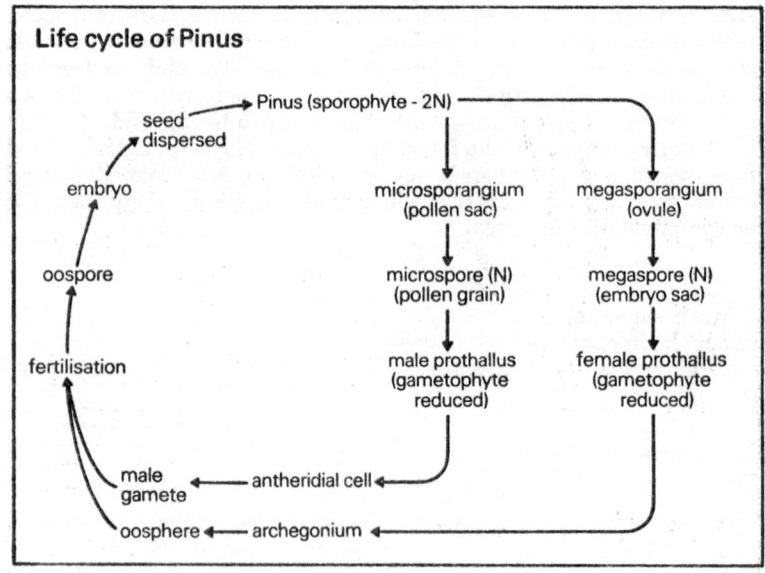

Pollination

This takes place in May. The **microsporangia** or pollen sacs split open along a median line liberating clouds of yellow pollen. It is non motile, and is dispersed by the wind, much of it being wasted. Some of the microspores settle between the scales of the **female cone**. Mucilage is secreted through the micropyle to trap the microspore. It dries and draws the microspore into the micropyle to rest near the apex of the nucellus. The scale closes, the cones bend back amongst the previous year's foliage. Fertilisation takes place next year.

Fertilisation

The **microspore** germinates. The vegative cell protrudes and elongates to form a pollen tube. The **antheridial cell** divides into a **static cell** and a **generative cell.** All pass down to the apex of the pollen tube. The naked generative cell divides into two **male gametes**. The pollen tube enters the **archegonium**. One **nucleus** fuses with an **oosphere** to form an **oospore**. The cone increases in size and becomes green. It is found at the apices of the shoots, below the terminal winter bud.

SPERMATOPHYTA

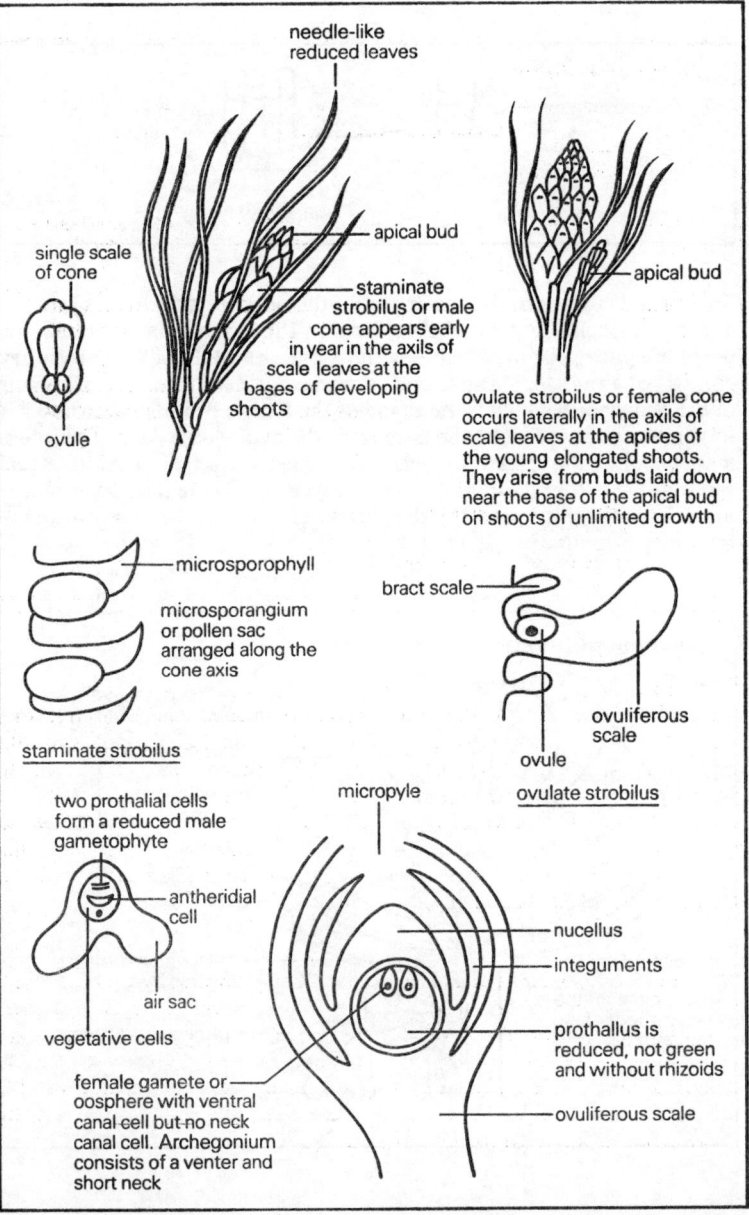

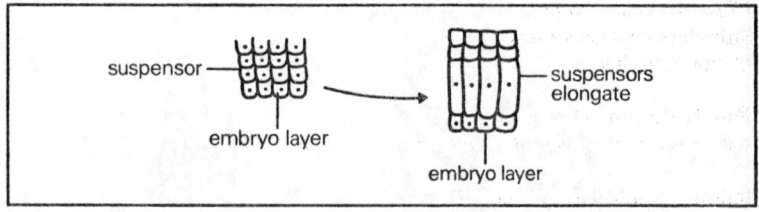

The nucleus passes to the lower end of the oospore and divides into four and eight nuclei. **Four embryos** form in Pinus from one oospore, i.e. **polyembryony**, but only one develops and the others die. The embryo consists of a **radicle, plumule** and **several cotyledons** formed in the centre of the **endosperm** which is the tissue of the **female prothallus**, stored with food material supplied by the parent via the **ovuliferous scale**. The **integument** becomes the **testa** of the seed. The upper surface of ovuliferous scale separates as a thin membrane to aid dispersal. The female cone reaches maturity in the third year. It is dry, brown and woody. Seeds are dispersed by the hygroscopic property of the scales.

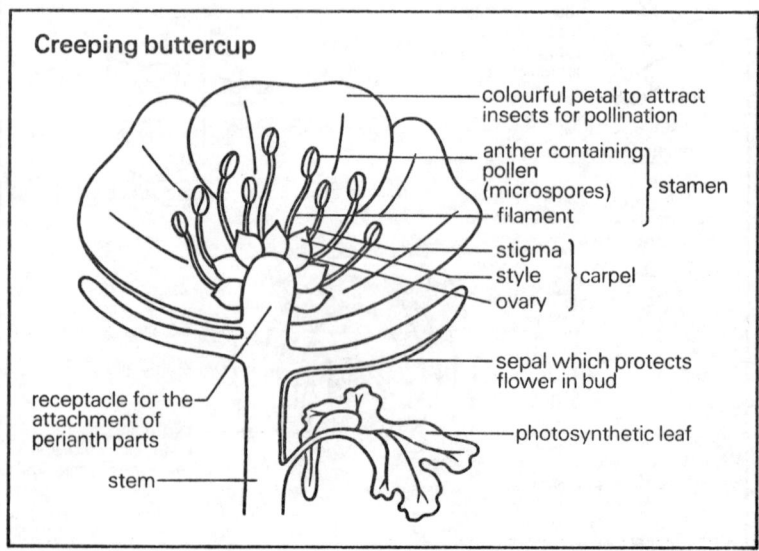

Class Angiospermae
Sub-class Dicotyledones
Order Ranales

Family Ranunculaceae
e.g. Ranunculus repens – creeping buttercup

Family Hippocastanaceae
e.g. Aesculus hippocastanum – horse chestnut

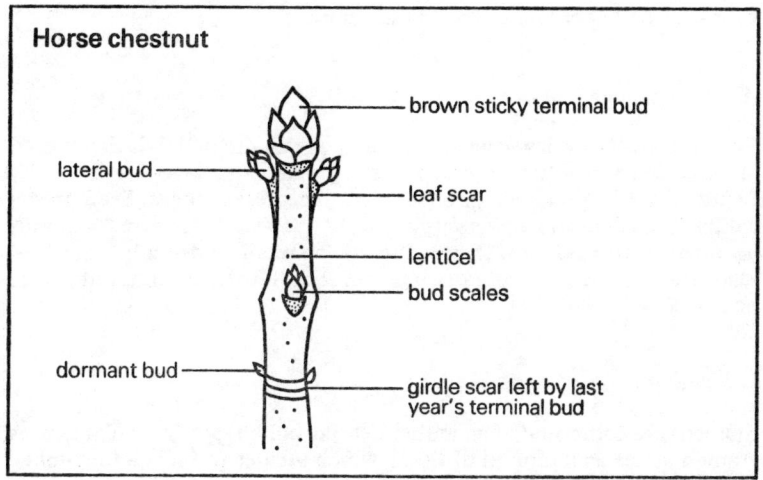

Annual – A plant that completes its life cycle in one year's growth e.g. Lathyrus sylvestris – Everlasting pea.
Biennial – A plant that requires two years to complete its life cycle. During the first year food is stored which is used in the second year e.g. Daucus carota – wild carrot.
Ephemeral – A plant with a life cycle so short that it completes several life cycles in a year e.g. Senecio vulgaris – Groundsel.
Perennial – A plant that continues its growth from year to year. Aerial parts of the herbaceous perennials die before winter and new shoots grow the following spring e.g. Ranunculus repens – creeping buttercup. Woody perennials are permanent woody stems above the ground. Many are de-

ciduous in the winter months and resume growth the following spring e.g. Acer campestre -- English Maple.

Vegetative propagation

This is a method of asexual reproduction in plants by some part of the plant body and its development into a new plant, e.g. bulbis in the lily, runners in the strawberry, offsets in the houseleek, stolons in the blackberry, suckers in the mint, rhizomes in the iris, bulbs in the onion, corms in the crocus, stem tubers in the potato, root tubers in dahlia and turions which are detached winter buds in saggitaria.

Artificial propagation

Cuttings, graftings, layering and budding are all methods which are used in the cultivation of fruit trees, roses etc. Some plants have lost the ability to propagate by seeding e.g. bananas and seedless oranges. Seed propagation takes time and yields varying results. It is easier to care for maturing plants than seedlings. Horticulturists can retain and multiply a special feature e.g. colour of a rose or quality of fruit. Flowering and fruiting can be controlled.

Pollination

Stamens are composed of an anther bearing pollen sacs and a filament. A stamen arises as a mound of tissue which divides to form a four-lobed anther. A pollen sac is found in each lobe. A single cell divides into two. One cell forms archesporial tissue and the other the tapetal cell. The tapetal cell divides to form three layers around the archesporium. The archesporial cells separate and function as micro-spore mother cells. Each divides by meiosis to form four haploid microspores or pollen grains which lie freely in the pollen sac.

When the outer fibrous layer of the anther shrinks the thinner-walled cells between adjacent pollen sacs rupture and the pollen grains are exposed. Before the pollen grains are set free they germinate. There are no prothalial cells. A two-layered microspore encloses a nucleus which divides into a generative and a pollen tube nucleus. Pollination involves the transference of pollen from the anthers to the stigma of the same flower in self-pollination, or to the stigma of another flower of the same kind in cross-pollination. Agents of pollination are the wind, insects and water. There are special mechanisms for cross-pollination. The advantage of cross-pollination is that it provides a greater variety of offspring.

Methods to ensure cross pollination

These are the mechanisms in flowering plants which prevent inbreeding.

Incompatibility

The genes in the pollen grain and those in the stigma, interact to make pollination by the same plant, or a genetically identical one, impossible.

For example the pin-eyed and thrum-eyed Primula vulgaris, the primrose.

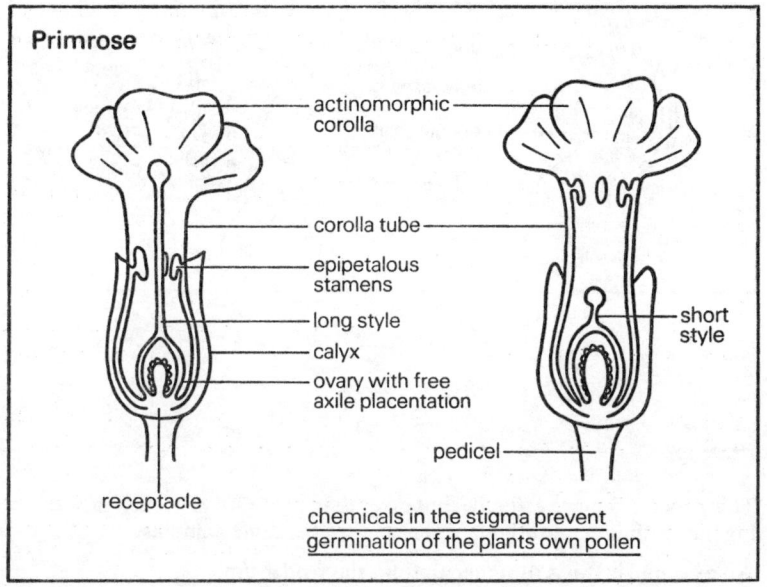

Protrandry

Self-pollination in these flowering plants is prevented by the ripening of the male stamens before the female stigma are mature.

The Lamium album, the white dead nettle, is an example.

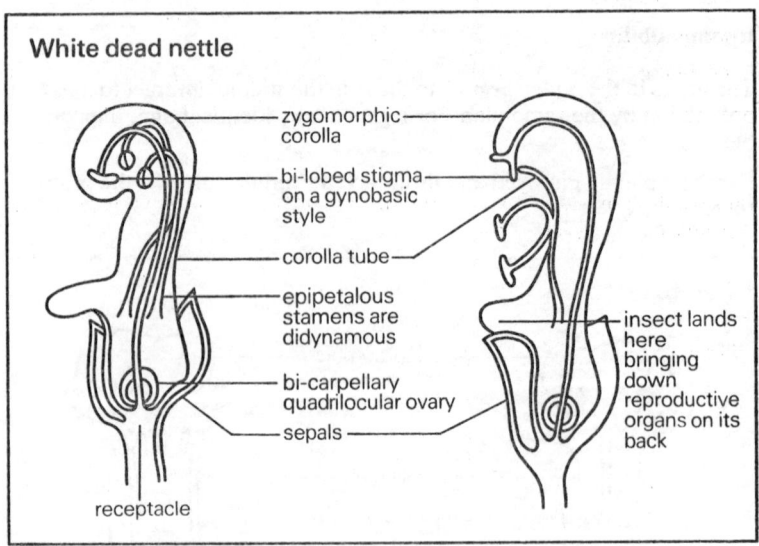

Protogyny

This process is more effective but rarer than protandry. In such flowering plants the female stigma ripens before the male stamens.

An example is the Arum maculatum. the wild arum.

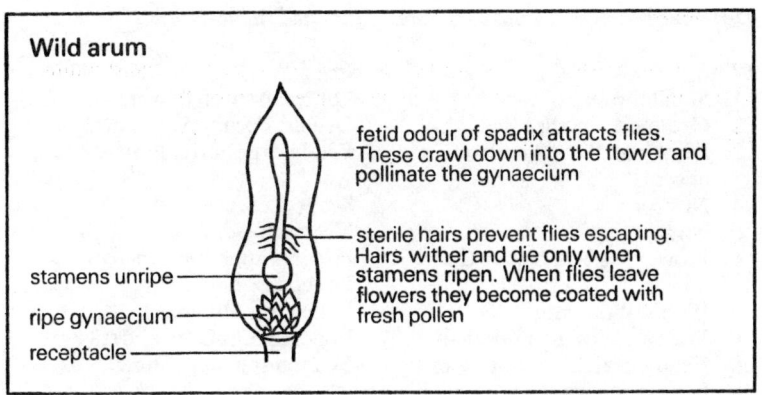

Unisexual plants e.g. Ilex aquifolium, the holly, or **unisexual flowers** e.g. Corylus avellana, the hazel.

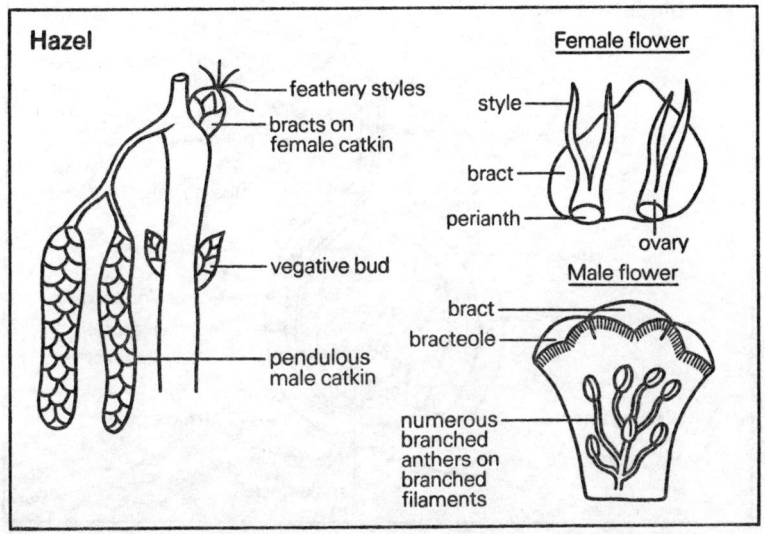

Differences between wind and insect pollinated flowers

Wind
1. Small flowers
2. Grouped together
3. Petals small and green or absent
4. No scent
5. No nectar
6. Flowers appear before the leaves
7. Pollen abundant
8. Pollen grains dry and small
9. Filaments long and anthers hang in the wind
10. Stigmas long and occasionally feathery

Insect
Large coloured flowers
Usually occur on their own
Petals large and colourful

Scent present
Nectar present
Flowers appear when many insects abound
Pollen less abundant
Pollen grains large and sticky
Filaments usually short

Stigmas short and sticky

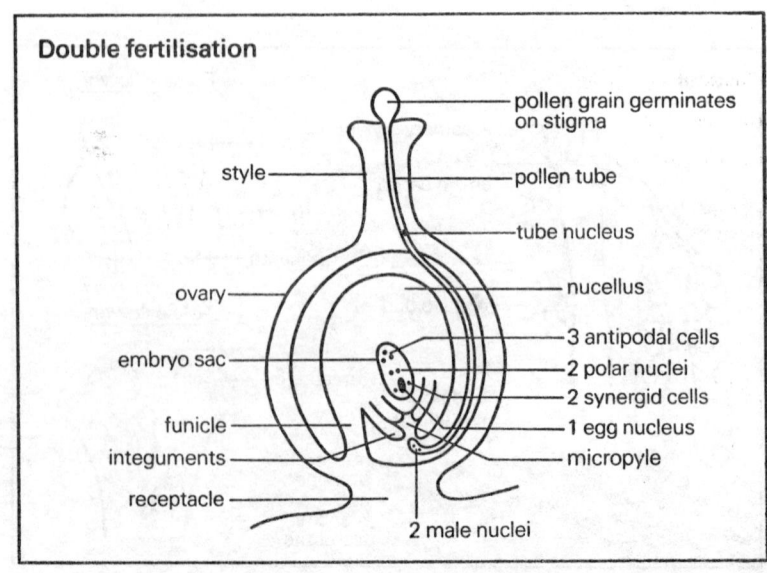

Double fertilisation

Fertilisation

The pollen deposited on the stigma germinates in a sugary substance produced from the stigma. A pollen tube grows down the style. The tip of the pollen tube grows into the micropyle and the two male nuclei contained in the pollen tube are liberated in the ovule. One nucleus fuses with the egg nucleus to form an oospore. The other nucleus fuses with the food nucleus. **Double fertilisation occurs.**

When fertilisation is complete, the flowering parts die and fall off. The wall of the ovary is called the pericarp and in the buttercup this becomes leathery. The ovule forms the seed which contains the embryo plant. The endosperm cell divides to form the endosperm. This grows by digesting the nucellar tissue which disappears as it is absorbed by the cotyledons of the embryo. The ovary thus forms a fruit which becomes dispersed before the seeds are liberated to germinate into a new plant.

Dispersal of fruits and seeds

1 **Wind**	seed parachutes	e.g. willow herb	– Epilobium hirsutum
	fruit parachutes	e.g. dandelion	– Taraxacum officinale
	winged seeds	e.g. gladiolus	– Gladiolus primulinus
	winged fruits	e.g. sycamore	– Acer pseudoplatanus
	separation of carpels	e.g. buttercup	– Ranunculus repens
	bracts	e.g. lime	– Tilia europaea
	censor mechanism	e.g. poppy capsule	– Papaver rhoeas
2 **Water**	Spongy aril	e.g. water lily	– Nymphaea alba
	fibrous exocarp	e.g. coconut	– Cocos nucifera
3 **Animals**	succulent seeds	e.g. yew	– Taxus baccata
	succulent fruits	e.g. blackberry	– Rubus fruticosus
	hooked fruits	e.g. cleavers	– Galium aparine
	hooked seeds	e.g. avens	– Geum urbanum
	oily seeds	e.g. gorse	– Ulex europaeus
4 **Explosive**	legume	e.g. pea	– Lathyrus sylvestris
	siliquas	e.g. stock	– Matthiola incana
	siliculas	e.g. honesty	– Lunaria annua

Germination
Germination occurs in favourable conditions i.e. suitable temperature, oxygen and water.

Hypogeal germination occurs when the cotyledons remain beneath the surface of the soil e.g. broad bean, Vicia faba. The plumule is protected during its passage through the soil by its own curvature.

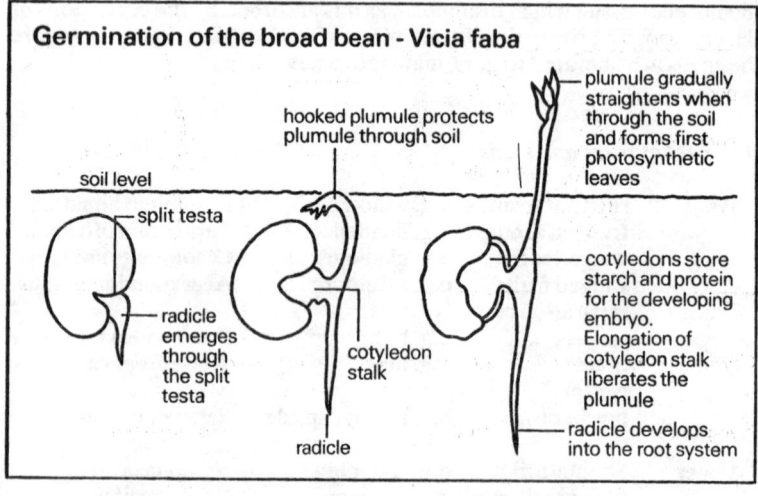

Epigeal germination occurs when the cotyledons come above the surface of the soil. The plumule is protected during its passage through the soil by two cotyledons and the curvature of the hypocotyl.

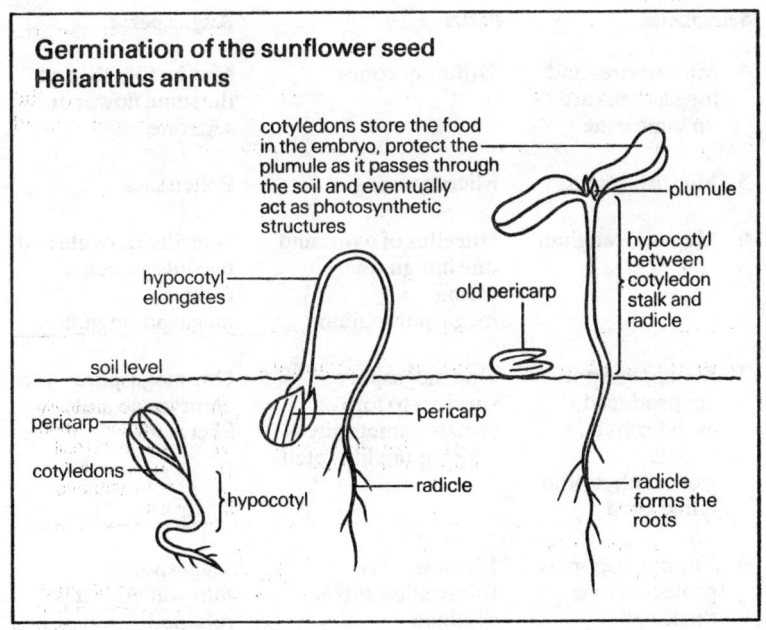

Comparisons between cycad, conifer and angiosperm

Selaginella	**Pinus**	**Angiosperm**
1. Alternation of generations	Alternation of generations	Alternation of generations
2. Sporophyte is dominant	Sporophyte is dominant	Sporophyte is dominant
3. Gametophyte generation is small, green and uses the stored food provided by the sporophyte	Gametophyte generation is much reduced and not independent	Gametophyte generation is much reduced and not independent

Selaginella	Pinus	Angiosperm
4. Microspores and megaspores are on same cone	Different cones	May be together on the same flower or separate
5. Microsporangia	Microsporangia	Pollen sacs
6. Megasporangium	Nucellus of ovule and one integument forms megasporangium	Nucellus of ovule and two integuments form megasporangium
7. Four megaspores are produced, each forms a female gametophyte and is liberated	One megaspore survives to form a female gametophyte which is not liberated	One megaspore – the embryo sac, not liberated
8. The megaspore is protected by a thick wall	Megaspore is thin-walled as it is retained	Megaspore is thin-walled as it is retained
9. Microspore cuts off a few prothallial cells and an antheridium containing motile spermatazoids	Microspore cuts of two prothallial cells and an antheridial cell	Microspore cuts off one vegetative cell. Two gametes are formed
10. Motile	Non-motile	Non-motile
11. No pollen tube	Pollen tube formed	Pollen tube formed
12. Megaspore forms two or three archegonia with a short neck with neck canal cells	Megaspore forms two or three archegonia with a short neck without neck canal cells	Archegonia are naked egg cells

Selaginella	Pinus	Angiosperm
13. Fertilisation by free swimming spermatozoids into an exposed archegonia	Fertilisation by non-motile gametes deposited in a pollen tube at neck of archegonia by wind	Fertilisation by non-motile gametes. Pollination agents involved in depositing pollen at female gametophyte. Pollen tube grows
14. No endosperm	Embryo becomes embedded in the endosperm and is the remains of a female prothallus, rich in stored food. It is the perisperm	Endosperm formed as a result of double fertilisation

Class Angiospermae

Sub-class Dicotyledones
Order Rhoedales
Family Cruciferae
e.g. Cheiranthus cheiri, the wall-flower.

General characteristics

1. These are herbaceous plants with simple exstipulate leaves, spirally arranged.
2. The inflorescence is a corymb without bracts.
3. The flowers are actinomorphic, hermaphrodite and hypogynous.
4. The calyx is composed of four free sepals arranged in two whorls. The inner pair may contain nectaries.
5. The corolla is composed of four claw-shaped petals alternating with the sepals in a cruciform arrangement.
6. The androecium is tetradynamous and is composed of six stamens, four long inner ones and two short outer stamens. All are introrse.
7. The gynaecium is composed of two carpels. They are syncarpous with parietal placentation. They are superior. A false septum occurs across the ovary.
8. The fruit is a siliqua, silicula or lomentum.
9. Pollination is by insects but often self-pollination occurs.
10. Examples are Capsella bursa-pastoria, Shepherd's purse, and Lunaria annua, Honesty.

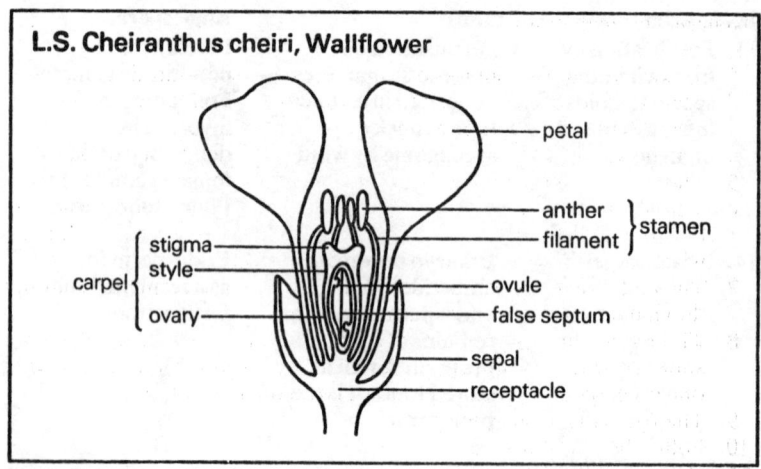

Floral formula
⊕ K4 C4 A2+4 G2

Floral diagram

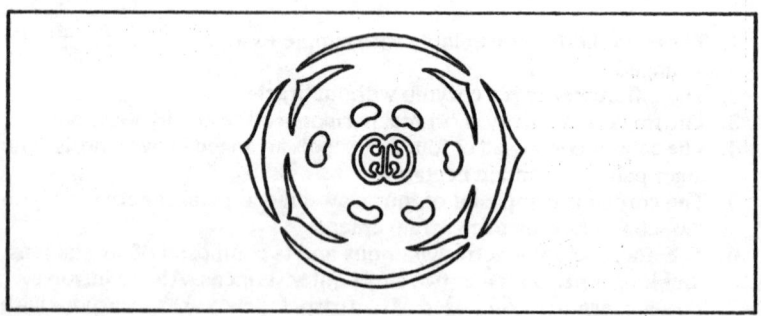

Sub-class Dicotyledones
Order Tubiflorae
Family Labiatae
e.g. Lamium album – White dead nettle.

General characteristics

1. These plants are widely distributed, herbaceous or shrubby.
2. They have square stems with opposite and decussate leaves.
3. The inflorescence is verticillaster or solitary.
4. The flowers are hermaphrodite, zygomorphic and hypogynous.
5. The calyx is composed of five joined sepals.
6. The corolla is composed of five petals joined to form a tube. They are bi-labiate with two posterior and three anterior petals, or labiate when the two upper petals are reduced in size.
7. The androecium is composed of four epi-petalous stamens. They are didynamous and alternate with the corolla petals.
8. The gynaecium is composed of two syncarpous carpels. They are superior with a gynobasic style. The ovary is divided into four with one erect ovule in each cell. They have axile placentation
9. The fruit is schizocarpic.
10. Pollination is by insects.
11. Examples are Glechoma hederacea – Groundy ivy, Ajuga reptans – Bugle.

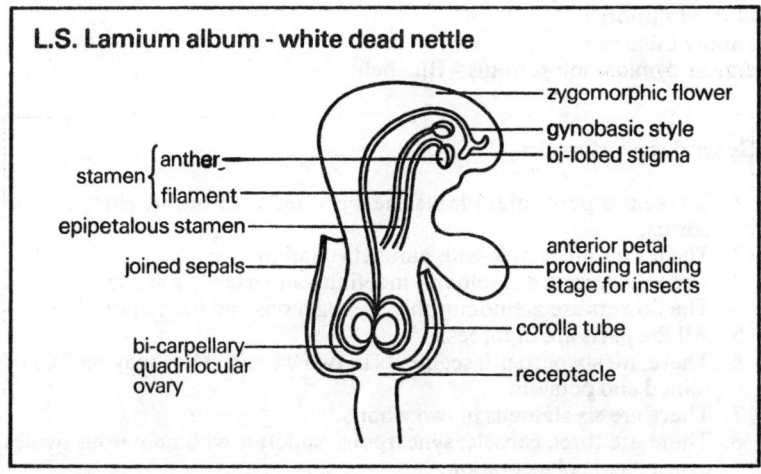

Floral formula

i.e. 1.K(5) $\overparen{C(5) A4}$ $\underline{G2}$

Floral diagram

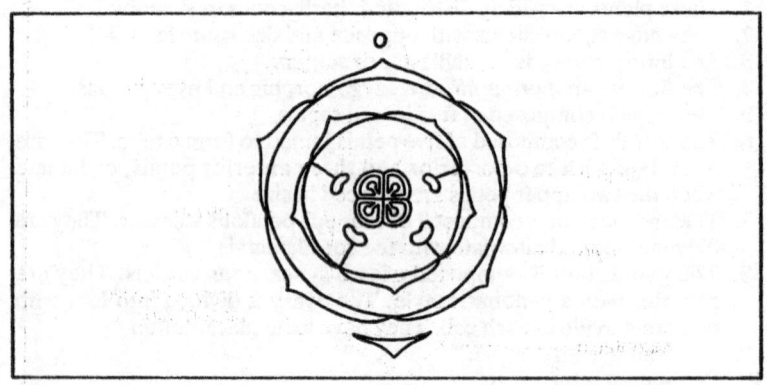

Sub-Class Monocotyledones
Order Liliiflorae
Family Liliaceae
e.g. Endymion non-scriptus – Bluebell

General characteristics

1. These are perennial plants, herbaceous with bulbs, rhizomes or corms.
2. The leaves are simple with parallel venation.
3. The inflorescence is solitary, indefinite and usually raceme.
4. The flowers are actinomorphic, hypogynous and hermaphrodite.
5. All the parts are in threes.
6. There are six perianth segments in two whorls. They may be free or joined and petaloid.
7. There are six stamens in two whorls.
8. There are three carpels; syncarpous, superior with numerous ovules or with axile placentation.
9. The fruit is a capsule or berry.
10. Pollination is by insects.
11. Examples are Allium oleraceum – Field garlic, Convallaria majalis – Lily-of-the-valley.

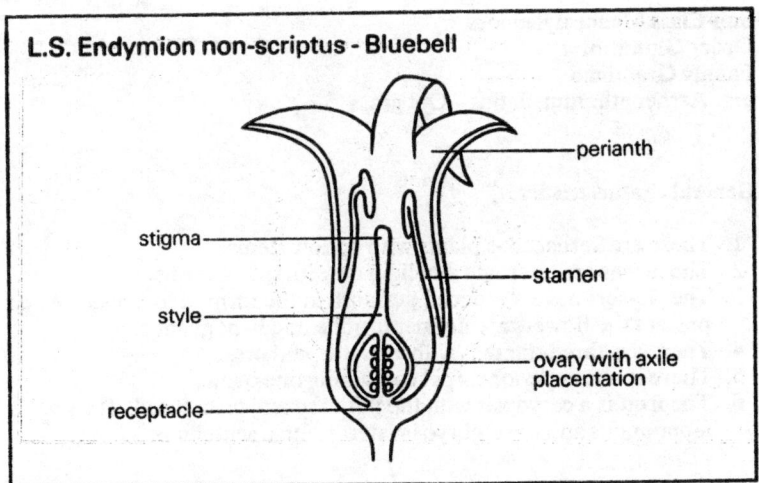

Floral formula

i.e. ⊕P(3+3) A3+3 G(3)

Floral diagram

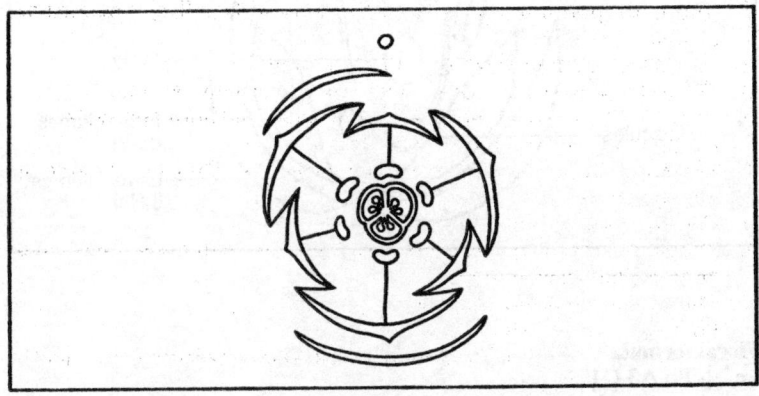

Sub-Class Monocotyledones
Order Glumiflorae
Family Graminae
e.g. Arrhenatherum elatius – Oat grass

General characteristics

1. These are herbaceous plants with hollow stems.
2. The leaves are alternate and ligulate with split sheaths.
3. The flowers have a reduced perianth in the form of lodicules. When present the flowers are hermaphrodite and hypogynous.
4. There are three stamens with versatile anthers.
5. There is one superior carpel containing one ovule.
6. The fruit is a caryopsis with the pericarp and testa fused. The seed is albuminous and the embryo is lateral with a scutellum.

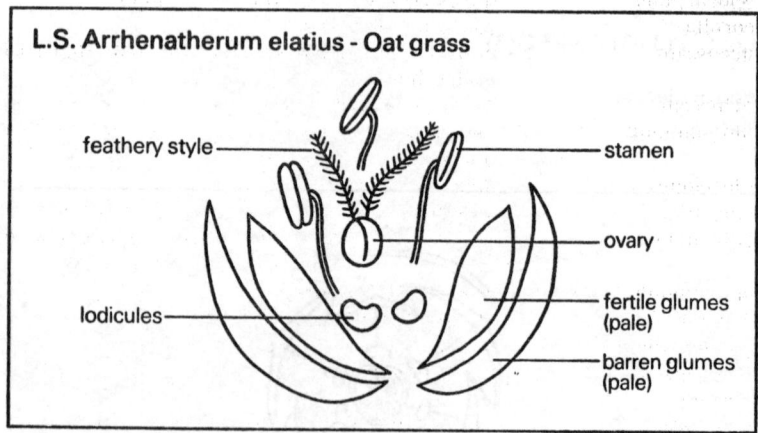

Floral formula
i.e. ·|· Po A3 G1

Floral diagram

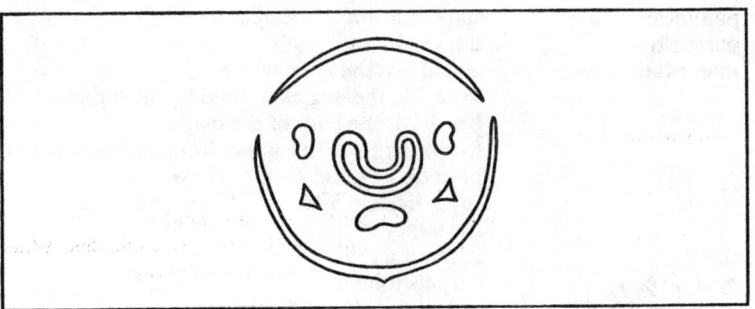

General terms

actinomorphic	having more than one plane of symmetry
alternate	arranged in two rows but not opposite
apocarpous	the carpels are free from one another
corolla	collection of petals
decussate	opposite with successive pairs at right angles to each other
dehiscent	opening to shed its contents
didynamous	two similar parts attached to one another by a portion of their surface
dioecious	male and female flowers
epicalyx	a calyx-like structure outside the real calyx
epigynous	an inferior ovary – floral parts attached above ovary
epipetalous	attached to the corolla
exstipulate	without stipules
gamopetalous	petals joined to form a tube
glabrous	without hairs
glaucous	bluish
gynobasic	a style which appears to be inserted at the base of an ovary
gynaecium	female part of the flower
hypogynous	a superior ovary – floral parts attached below ovary
inflorescence	flowering branch

introrse	opening towards the middle of the flower
pedicel	stalk of a single flower
peduncle	stalk of a single inflorescence
perianth	the sepals and petals
placentation	apical – at the apex of the ovary
	axile – in the angles formed by the septum
	basal – at the base of the ovary
	free central – on a column from the base of ovary
	parietal – on the walls of the ovary
	superficial – ovules scattered over the walls of the ovary
sessile	without a stalk
synocarpous	carpels joined
zygomorphic	only one plane of symmetry

Practice Questions

1. What are the general characteristics of the Spermatophyta?
2. What feature distinguishes the Gymnospermae from the Angiospermae?
3. Explain the term dioecious.
4. Distinguish between the male and female cones of Cycas.
5. The megaspore germinates to form the female gametophyte. Describe the female gametophyte.
6. Distinguish between the staminate and ovulate strobilus in Pinus.
7. Give an account of fertilisation in Pinus.
8. Make a large labelled diagram of the male gametophyte of Pinus.
9. Explain the term polyembryony.
10. What is the function of a) sepals, b) petals and c) receptacle?
11. Make a large labelled diagram of a two year old horse chestnut twig in winter.
12. With examples distinguish between an annual, biennial and perennial.
13. Write an essay on vegetative propagation.
14. What are the advantages of artificial propagation?
15. Distinguish between pollination and fertilisation.
16. Explain protandry and protogyny.
17. Tabulate the differences between wind and insect pollinated flowers.
18. What do you understand by double fertilisation?
19. Write an essay on the dispersal of fruits and seeds.
20. Distinguish between hypogeal and epigeal germination.
21. What is an endosperm?
22. Cheiranthus cheiri flowers are actinomorphic, hermaphrodite and hypogynous. Explain these terms.
23. Bugle, Ajuga reptans, is a member of the family Labiatae. What general characteristics would you expect it to have?
24. Distinguish between axile and free placentation.
25. Explain the floral formula $\oplus P(3+3)\ A3+3\ \underline{G\,(3)}$.

11 Climbers, xeromorphs, halophytes and hydrophytes

Climbing plants

When plants live crowded together they compete for light. This competition means that they will struggle to reach greater heights. The climbing habit enables plants to reach the light with minimum expenditure upon the stems by using an external support. Flowers of climbing plants are exposed for pollination and fruits for dispersal. Climbing plants may be divided into three groups a) straggling, b) prehensile and c) adhesive climbers.

a) Straggling
The rose rambler stem grows fast and is unable to support itself. The stem falls against a neighbouring support. The bent back hooks catch onto an object and hold the stem firmly.

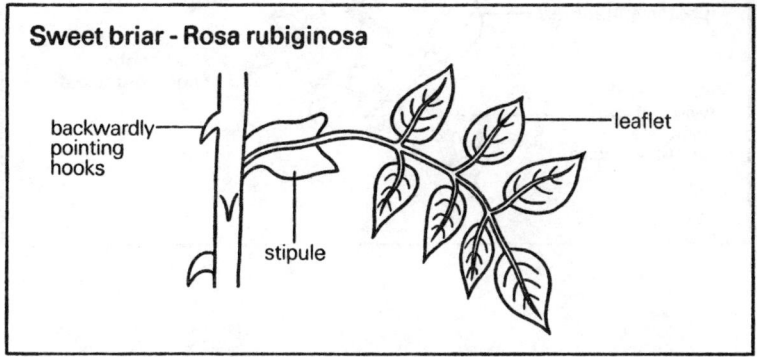

b) Prehensile climbers
The stems of prehensile climbers grow spirally around a support. The apical stem circumnutates. The twining elongates the internodes, delaying leaf production until a support is found. Twining is dextrorse e.g. honeysuckle – Lonicera periclymenum, or sinistrose e.g. Bindweed – Convolvulus arvensis.

Other plants climb by tendrils, formed from the tips of the laminae as extended parts of the lower region of the leaf, and have stipular or small shoots.

A tendril is a cylindrical, whip-like organ with a hooked tip which circumnutates in the air.

When the tendril comes in contact with a support, inequality of growth occurs causing the tendril to lap around the support. Once fixed the tendril strengthens its tissue and forms a spiral spring-like organ to help with support.

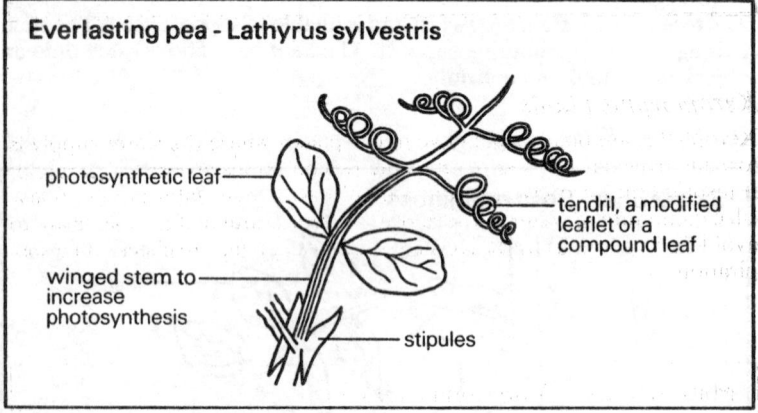

c) Adhesive climbers
Adhesive climbers attach themselves by close application of some part of the plant to a supporting surface. Ivy – Hedera helix, attaches itself by adventitious roots arising laterally from the stem. Virginia creeper – Parthenocissus benryana, has branched tendrils of which each branch expands into a disc which, with the aid of adhesive cement, forms an organ of attachment.

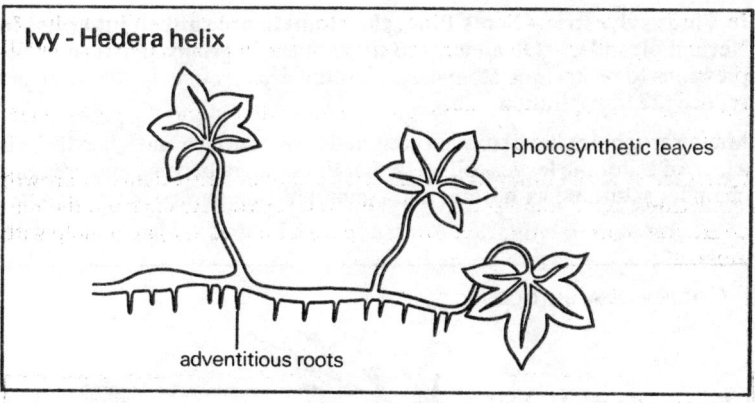

Xeromorphic plants

Xerophytes are plants which live in dry places where the water supply is low and irregular and where atmospheric conditions favour a high rate of transpiration. Xerophytes must economise with their water supply. A lavish expenditure of water in transpiration is undesirable if the plants are to avoid desiccation. The strange shapes of these plants are checks on transpiration.

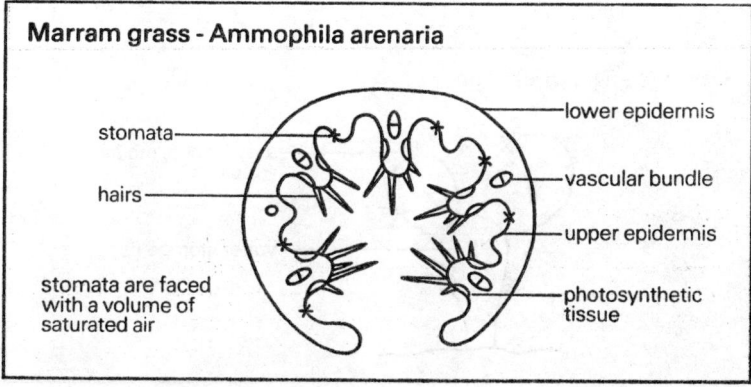

In Pinus sylvestris – Scots Pine, the stomata are sunken into pits. In Nerium oleander – Oleander, the stomata are in groups in irregular depressions to protect the stomata against the drying effect of the wind and reduce the transpiration rate.

Many xerophytes have stomata confined to one surface and when there is a lack of water the leaves roll up so that the stomata become enclosed in a chamber saturated with water e.g. Ammophila arenaria – Marram grass.

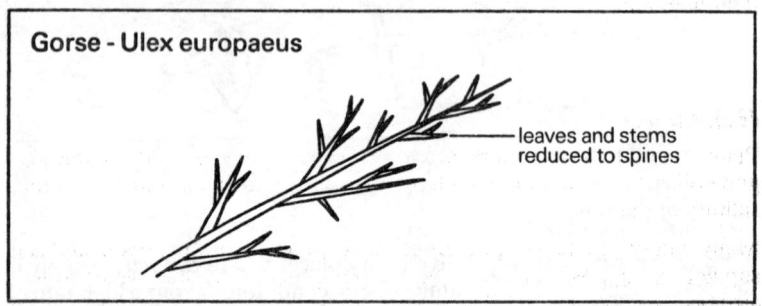

Some Labiates have a covering of hairs. Other xerophytes reduce the number of leaves by forming leaf spines e.g. Gorse – Ulex europaeus, or have a small number of leaves e.g. Heather – Erica tetralix. Casuarina sheds its leaves.

Cuticular transpiration is prevented by a thick cuticle.

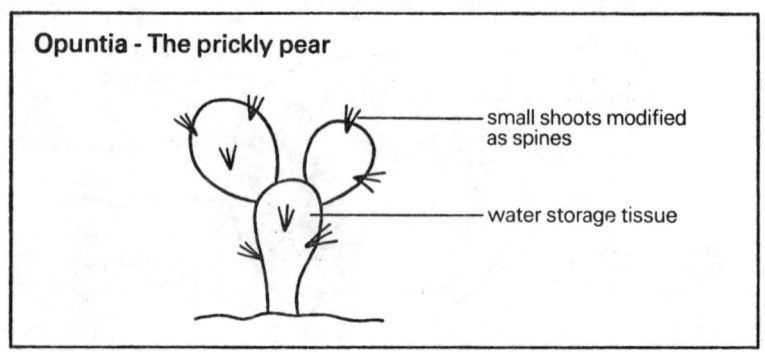

Some xerophytes develop water storage tissue with succulent stems. This is helped by large amounts of mucilage enabling the water to be held by imbibition e.g. Opuntia – the prickly pear. They have extensive root systems which cover large volumes of soil. A slow expenditure of water is due to few stomata protected by thick cuticles and mucilaginous contents of cells.

Excessive water loss can be prevented by a cessation of vegetative growth. Bulbous plants pass a drought period dormant, while deciduous trees lose their leaves to prevent excessive transpiration.

Halophytes

Plants that inhabit salt marshes face a) a lack of oxygen in the water logged soil, b) mobile mud which makes rooting difficult and c) extreme salinity of the soil.

Many halophytes have succulent stems with water storage tissue and cell sap with a high osmotic strength, enabling them to absorb water from a strong salt solution e.g. Glass wort – Salicornia europaea. Many have an extensive aerating system in which oxygen can be stored. The movement of mud is overcome by the presence of long rhizomes and roots which act as anchors in the loose mud.

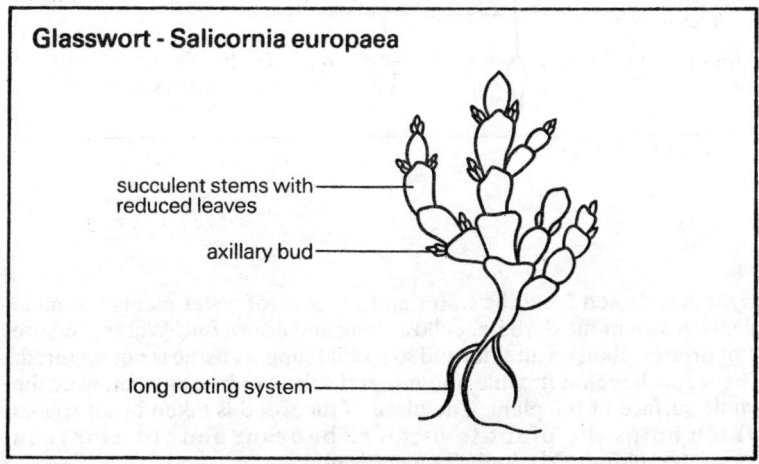

Glasswort - Salicornia europaea
- succulent stems with reduced leaves
- axillary bud
- long rooting system

Hydrophytes

Hydrophytes are plants found in lakes and ponds. They experience no extremes of temperature, little danger of drying up and carbon dioxide is plentiful. However, light is poor and so limits photosynthesis. Hydrophytes, compared with other plants, show a reduction in all structures concerned with water. The xylem and cuticle are poorly developed with the foliage thin and delicate. A similar structure is found in submerged plants as the water provides sufficient support to the plant. Buoyancy and aeration present difficulties. In floating leaves the stomata are found on the undersurface and the internal air spaces are greatly enlarged to ensure adequate aeration and buoyancy. Many changes occur in the stem, root and leaves.

Roots tend to disappear, e.g. duckweed – Lemna polyrrhiza, and those that are retained are used to fix the plant in the mud and have no root hairs.

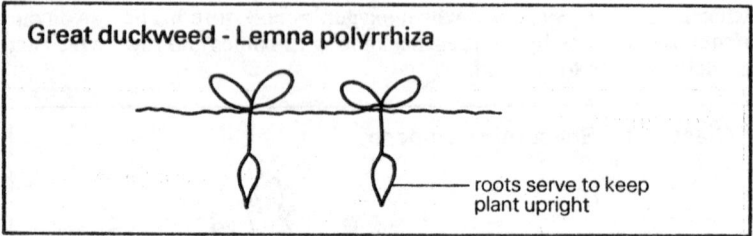

Light is reflected from the water and so stems of water plants resemble plants grown in the dark i.e. yellow, long and drawn out. Water pressure supports the plant on all sides and so special support tissue is not required. The stems become flexible and mineral salts can be absorbed over the whole surface of the plant. The place of the wood is taken by air spaces which helps the plant to be more buoyant and store oxygen e.g. water milfoil – Myriophyllum verticallatum.

CLIMBERS, XEROMORPHS, HALOPHYTES AND HYDROPHYTES

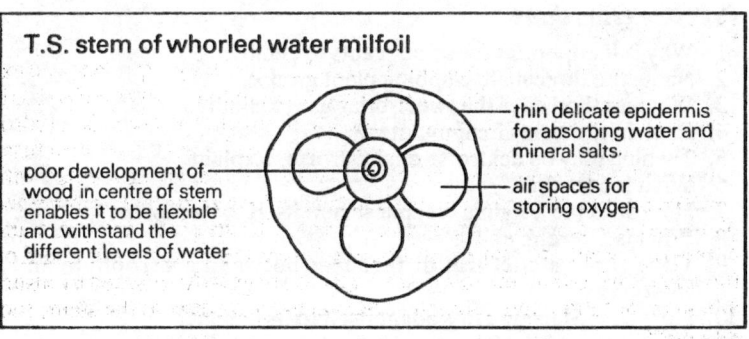

Three types of leaves have developed in connection with the adoption of water habit e.g. Arrowhead – Sagittaria sagittifolia L.

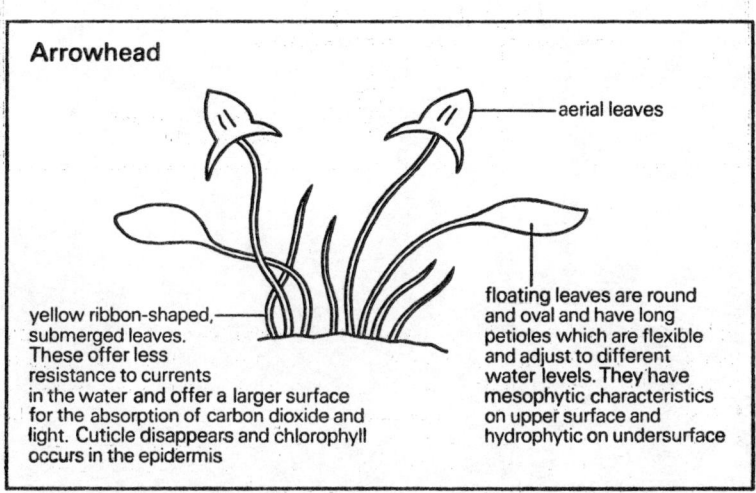

Practice Questions

1. Why is it important for some plants to climb?
2. Name the three main climbing plant groups.
3. What method does the sweet-briar use to climb?
4. Explain the term circumnutates.
5. Twining may be dextrorse or sinistrorse. Explain.
6. What is a tendril?
7. How does the Virginia creeper secure itself to its support?
8. What is a xerophyte?
9. Give three structural differences between a xerophyte and a mesophyte.
10. How does Marram grass adapt itself to its habitat?
11. Describe a habitat where you would expect to find Ulex europaeus.
12. What difficulties do halophytes face and how do they overcome these difficulties?
13. Name three hydrophytes.
14. What reduction in structure do hydrophytes show and why?
15. Describe the leaves found in Saggitaria and give reasons for their differences.

NOTES

NOTES

NOTES

NOTES

NOTES

NOTES

NOTES

IMPROVE YOUR
CELTIC REVI

In a tough world, every qualification counts towards a brighter future. So naturally, you want to do all you can to get through those exams.

And now, with *Celtic Revision Aids,* you can really do something about it.

By choosing titles from those series in the Celtic range which are appropriate to you, you can plan long-term course preparation, or carry out last minute revision. You can study model answers and essay guidelines, or check your knowledge against lists of basic facts.

Celtic Revision Aids are a complete range of books, covering subjects both in depth and on an 'essential fact' level.

In short, they're designed to make revision easier, designed to help you to examination success.

Series and titles available:

Rapid Revision Notes O-level
£1.50 each
Titles: *English Language Book 1, English Language Book 2, Mathematics, Physics, Chemistry, Biology, Human Biology, Physical Geography, Commerce, Economics, Sociology, British Economic History, Integrated Science, Commercial Mathematics, Accounts.*

Rapid Revision Notes A-level
£1.75 each
Titles: *Pure Mathematics, Applied Mathematics, Statistics, General Biology, Botany, Zoology, Inorganic Chemistry, Organic Chemistry, Physical Chemistry, Physics – Mechanics, Physics – Heat, Light and Sound, Physics – Electricity and Magnetism.*

EXAM RESULTS
REVISION AIDS

Literature Revision Notes and Examples
£1.50 each
Titles: *Chaucer's Prologue, Merchant of Venice, Julius Caesar, Richard II, A Midsummer Night's Dream.*

New Testament Studies
£1.50 each
Titles: *St. Matthew, St. Mark, St. Luke, St. John, Acts of the Apostles.*

Law Revision Notes
£1.95
Titles: *Principles of Law, Criminal Law, Family Law, Law of Tort, Company Law.*

Model Answers
£1.25 each
Titles: *Julius Caesar, Macbeth, Romeo & Juliet, Merchant of Venice, Aids to Mathematics, Aids to Essay Writing, English Language, Model Essays, Practice in Summary Writing, Essay Plans, Biology, Chemistry, Human Biology, Physics, Mathematics, Commerce, Economics, British Isles Geography, British Economic History, Accounts, Commercial Mathematics, Integrated Science.*

Test Yourself
95p each
Titles: *English Language 1, English Language 2, French, German, Commerce, Economics, Chemistry, Physics, Mathematics, Modern Mathematics, Biology, Human Biology, St. Matthew, St. Mark, St. Luke, St. John, Acts of the Apostles, Commercial Mathematics, Accounts, Statistics, British Isles Geography, British Economic History.*

Multiple Choice O-level
£1.25 each
Titles: *English 1, English 2, French, Mathematics, Modern Mathematics, Chemistry, Physics, Biology, Human Biology, Commerce, Economics, British Isles Geography, Accounts, Commercial Mathematics, Integrated Science.*

Multiple Choice A-level
£1.50 each
Titles: *Pure Mathematics, Applied Mathematics, Chemistry, Physics, Biology, Statistics.*

Worked Examples A-level
£1.50 each
Titles: *Pure Mathematics, Applied Mathematics, Chemistry, Physics, Biology, Economics 1, Economics 2, Sociology, British History 1914-76, European History 1914-76, British Economic History, Physical Geography, Accounts, Statistics.*

For a full colour brochure giving details of the complete range of Celtic Revision Aids, send your name and address to:

Celtic Revision Aids – Direct Brochure Mailing, Dept. T.,
TBL Book Service Ltd.,
17-23 Nelson Way,
Tuscam Trading Estate,
Camberley,
Surrey.
GU15 3EU

Celtic Revision Aids are available from good booksellers everywhere or, in cases of difficulty in obtaining them, direct from the publisher. If you wish to order books direct from Celtic, write to the above address giving your own name and address in block capitals, clearly stating the title/s of the book/s you would like, and enclosing a cheque or postal order made payable to TBL Book Service Ltd., to the value of the cover price of the books required *plus* 25p postage and packing per order for orders up to 4 books. (Postage and packing is free for orders of 5 books or more.) (U.K. only.)

Celtic Revision Aids reserve the right to show new retail prices on covers which may differ from those previously advertised in the text or elsewhere, and to increase postal rates in accordance with the P.O.